PRENTICE HALL'S
ENVIRONMENTAL TECHNOLOGY SERIES

Volume 4

Sampling and Analysis

Prentice Hall's Environmental Technology Series

Available:

Volume 1: Introduction to Environmental Technology

Volume 2: Environmental Regulations Overview

Volume 3: Health Effects of Hazardous Materials

Volume 4: Sampling and Analysis

Volume 5: Waste Management Concepts

Planned:

Volume 6: Industrial Waste Stream Generation

PRENTICE HALL'S
ENVIRONMENTAL TECHNOLOGY SERIES

Volume 4

Sampling and Analysis

NEAL K. OSTLER, Editor
Salt Lake Community College

PATRICK K. HOLLEY, Editor

Prentice Hall
Upper Saddle River, New Jersey   *Columbus, Ohio*

Library of Congress Cataloging-in-Publication Data

Ostler, Neal K.
Sampling and analysis / Neal K. Ostler,
Patrick Holley.
p. cm. -- (Prentice Hall's environmental technology series;
v. 4)
Includes bibliographical references and index.
ISBN 0-02-389534-9
1. Pollution -- Measurement. 2. Environmental sampling.
I. Holley, Patrick II. Title. III. Series.
TD193.088 1997
628.5'028'7--dc21 96-48897
CIP

Editor: Stephen Helba
Production Editor: Mary Harlan
Design Coordinator: Julia Zonneveld Van Hook
Text Designer: Custom Editorial Productions, Inc.
Cover Designer: Brian Deep
Production Manager: Pamela D. Bennett
Marketing Manager: Frank Mortimer, Jr.
Editorial/Production Supervision: Custom Editorial Productions, Inc.
Illustrations: Custom Editorial Productions, Inc.

This book was set in Utopia by Custom Editorial Productions, Inc., and was printed and bound by Quebecor Printing/Book Press. The cover was printed by Quebecor Printing/Book Press.

A Pearson Education Company
Upper Saddle River, NJ 07458

Printed in the United States of America

10 9 8 7 6 5 4 3 2 1

ISBN 0-02-389534-9

Prentice-Hall International (UK) Limited, London
Prentice-Hall of Australia Pty. Limited, Sydney
Prentice-Hall Canada Inc., Toronto
Prentice-Hall Hispanoamericana, S.A., Mexico
Prentice-Hall of India Private Limited, New Delhi
Prentice-Hall of Japan, Inc., Tokyo
Pearson Education Asia Pte. Ltd., Singapore
Editoria Prentice-Hall do Brasil, Ltda., Rio De Janeiro

Contents

Preface

This text has been prepared as a practical guide to sampling and analysis techniques, which are used commonly in the environmental and materials testing fields. Many contaminated sites around the country are undergoing site evaluation and characterization, requiring the extensive use of sampling and analysis programs. Laboratories are performing analytical procedures to determine levels of chemical contamination in air, water, soil, and waste materials, as well as the physical properties of materials. This text is not intended to be an all-inclusive, but it does provide technical information on the topics of concern to those who are enrolled in college-level technical environmental programs, engaged in training for current workplace needs, or interested in the field.

This text is designed to impart the knowledge and skills necessary to understand sampling and analysis concepts and to conduct basic sampling procedures. The text discusses sampling plans; air, water, soil, and waste sampling procedures and techniques; data interpretation; sample handling and preservation techniques; nondestructive materials testing techniques; analytical laboratory sample handling practices; health and safety considerations; quality assurance and control procedures; and documentation requirements.

Sampling and analysis techniques may be best presented by utilizing a combination of instructional methods including lectures, slides and overheads, classroom exercises, field exercises, field trips, and classroom demonstrations. Materials are included in this text to assist in such presentations. The student should have access to as many sampling devices and technologies as possible to practice and become familiar with equipment, methods, and techniques. Included at the end of each chapter are additional materials, review questions, and activities for students to use.

Utilization of resources available within your local community college or university and the local business community is a very important component in presenting sampling and analysis techniques. Chemistry and environmental technology departments may provide assistance with chemical demonstrations, sampling equipment, and supplies. The geology department may provide assistance with hydrogeology, which includes subsurface sampling, groundwater well design, and groundwater sampling procedures. Laboratories or local environmental health departments may assist in providing information on analytical laboratory procedures and practices. These resources may also provide assistance with safety practices when applying sampling procedures on contaminated sites.

The authors of this text have prepared materials that are field practical and are based on detailed knowledge of current sampling and analysis procedures. The text represents a coordinated effort by the contributing authors to provide useful and timely information for use in technical training efforts. It has been our pleasure to work with all of the contributing authors and the publisher in compiling and coordinating the elements of this text. We wish to thank those who reviewed the manuscript, particularly Dr. Alison Morrison-Shetlar, Georgia Southern University.

Neal K. Ostler
Patrick K. Holley

About the Authors

Neal K. Ostler

Mr. Ostler is an adjunct faculty member of Salt Lake Community College (SLCC), where he has been instrumental in implementing the associate degree program of Environmental Technology (ENVT). He has also developed the College's noncredit program, which the Utah Board of Regents recently recognized as the Environmental Training Center. He is a graduate of the University of Utah and has completed the ENVT program at SLCC. A Certified Hazardous Materials Manager, he has "Train-the-Trainer" certification in a variety of environmental health and safety (EH&S) subjects. He recently obtained the Designated Trainer for ISO 14000 Awareness credential from the Global Environmental Technology Foundation.

Mr. Ostler has a broad background including experience in law enforcement as a deputy sheriff and in hazardous materials emergency response. While working for an alcohol and substance abuse program at the Utah State Prison, he developed a bibliotherapy program for maximum security inmates. During his tenure as a motor carrier investigator with the Utah Department of Commerce, he began to develop his hazardous materials credentials and has now attended more than 1,000 hours of EH&S workshops and seminars.

In the spring of 1996, Mr. Ostler married Karen Amundsen of Park City, Utah. Their combined family includes ten children in their early teens and early twenties.

Patrick K. Holley

Mr. Holley is an environmental consultant with fourteen years of experience in laboratory, research, chemical plant, shipbuilding, and gas and electric utility operations. He is a Registered Environmental Assessor in California and a Certified Environmental Trainer. His background and experience include responsibilities in analytical laboratories, environmental compliance, remediation, and site assessment activities for Fortune 500 companies. In his role in the electric utility industry, Mr. Holley coordinated and maintained a groundwater monitoring system for a large electric generating plant. In his role in a research lab, he developed laboratory analytical techniques, collected samples, and calibrated and maintained laboratory analytical instruments. In his current role as an environmental consultant, he advises clients on regulatory compliance, sampling programs, and health and safety issues. Mr. Holley is a graduate of Mississippi State University, where he earned his B.S. degree in industrial technology.

Sim D. Lessley

Dr. Lessley is Executive Vice President of DataChem Laboratories with responsibility for laboratory operations. He has worked with the UBTL division of the University of Utah Research Institute as an analytical chemist in areas of gas chromatography, high performance liquid chromatography, X-ray diffraction, atomic absorption, and wet chemical analyses. He also served as Organic Chemistry group leader and Organic Section manager and as technical manager with responsibility for proposal preparation, project management, and problem solving. He was appointed Associate Director of UBTL, Inc., when the division became a for-profit corporation. In this new position, he was responsible for contract procurement, project management, and continued supervision of the Inorganic Chemistry department.

Dr. Lessley was appointed vice president of the newly formed DataChem Laboratories in 1987. His responsibilities at that point included oversight of the Department of Defense contract procurement efforts, supervision of project managers, and technical support of the sales and marketing group. In 1992, he assumed responsibility for the project managers' group and early in 1995 he was asked to oversee the Organic and Inorganic Departments at DataChem Laboratories. This responsibility involves coordinating staff and instrumentation to deliver analytical results according to standard written methods. Dr. Lessley received his Ph.D. in chemistry from the University of Utah.

Dean R. Lillquist

Dr. Lillquist is Assistant Professor in the industrial hygiene/hazardous substance program at the Rocky Mountain Center for Occupational and Environmental Health. Past professional experiences includes work with the OSHA–Consultation Program, interagency occupational and environmental health consultation with the U.S. Public Health Service, and the internal environmental protection and occupational health and safety programs of the National Institutes of Health in Bethesda, Maryland. Dr. Lillquist obtained his doctorate in environmental health (industrial hygiene/industrial toxicology) from Colorado State University and his

MSPH in environmental public health from the University of Minnesota. He is ABIH certified in the comprehensive practice of industrial hygiene.

Mark Sabolik

Mr. Sabolik is certified as a Level III in radiographic and ultrasonic testing by the American Society for Nondestructive Testing (ASNT) and as a Level II in eddy current, liquid penetrant, and magnetic particle testing. He is the chairman of the Robert J. Oliver Scholarship Committee of the ASNT and is a member of its Accreditation Council. He is the past chairman of the ASNT Great Salt Lake Section and is the current chairman of the local educational council.

Mr. Sabolik has worked for an independent testing laboratory and has tested weldments and components for more than 300 companies including aerospace, military and commercial aircraft, petrochemical, building construction, mining organizations, and the Department of Energy. His responsibilities at Salt Lake Community College include designing and building its nondestructive testing program and serving as an instructor and radiation safety officer. He also owns a corporation that writes procedures for specialized NDT applications, radioactive material licenses, and computer-based education programs. He is a graduate of Hutchinson Technical College.

David O. Wallace

Mr. Wallace is a clinical instructor in the industrial hygiene program at the Rocky Mountain Center for Occupational and Environmental Health and is the director of asbestos and lead training through the Center's continuing education program. His professional background includes broad-based industrial hygiene consulting and experience in the primary metals and chemical industry. Mr. Wallace obtained his MSPH in industrial hygiene from the University of Utah. He is ABIH certified in the comprehensive practice of industrial hygiene.

Michael A. Williams

Mr. Williams is a registered professional geologist with the State of Wyoming. He has thirteen years' experience as a hydrogeologist in the petroleum, mining, and environmental industries. He has eight years of experience conducting and supervising groundwater monitoring programs for RCRA, LUST, NPDES, and SDWA projects. Mr. Williams has prepared numerous sampling and analysis plans for sites contaminated with paint wastes, chlorinated solvents, petroleum hydrocarbons, and radionuclides. He is currently employed as a project hydrogeologist with Kleinfelder, a western geotechnical, environmental, and water resources engineering firm, in Salt Lake City, Utah. Mr. Williams earned his B.A. degree in geology and an M.S. degree in geology and water resources from Iowa State University in Ames, Iowa.

1

Overview of Sampling Operations

Patrick K. Holley

Upon completion of this chapter, you will be able to do the following:

- Identify and describe conditions that affect sample collection, preservation, and handling procedures.
- Understand and apply information contained in a sample collection, preservation, and laboratory method chart.
- Understand the concept of representative sample collection.
- Understand and apply different types of sample selection methods.
- Understand and describe basic units of measure used in sampling and laboratory work.
- Define basic terms associated with field sampling activities.

INTRODUCTION TO SAMPLING

Sampling is one of the most important components of environmental work. All decisions that characterize a hazardous material, a hazardous waste site, groundwater contamination, or water quality begin with proper sampling and subsequent chemical analysis. Substantial errors involving the incorrect identification and classification of materials on sites may be induced by the use of improper sampling techniques or procedures.

In addition to field sampling errors, laboratories also may contribute to errors when proper quality assurance and quality control procedures are not adhered to. An understanding of the chemical analysis and detection methods that are employed in lab instrumental methods also is key in selecting the correct sample collection, handling, and preservation techniques. This instructional guide will assist in the education of people who may be required to develop, manage, or conduct these field activities, but is not intended to be a complete procedural guide. Techniques, procedures, and concepts will be discussed to ensure the thorough understanding of the collection of various types of samples. Individuals engaged in sampling activities should always refer to current regulatory documents, site sampling plans, and laboratory guides prior to sampling efforts for exact sample collection procedures. It is important to note that exact procedures specified by the entity overseeing the sampling and/or the involved laboratory must be followed to ensure the integrity of any site sampling program. If the sample collection procedure specifies that the sample is to be preserved by cooling to a temperature of 4° Celsius, do it!

SAMPLING CONCEPTS

General considerations such as field practicality, quality assurance, program cost, and physical safety also should be incorporated into the sampling process. Laboratory analytical procedures are not the topic here but are addressed in EPA documents (which focus on laboratory analytical procedures yet also give insight into sample collection requirements) such as *Test Methods for Evaluating Solid Waste (SW-846)* (EPA, 1993) or *Standard Methods for the Examination of Water and Wastewater, 18th ed.* (EPA, 1992).

Collection of samples from the field requires knowledge and employment of concepts and techniques discussed in this text. Sources of sampling procedures may be found in the EPA's *Compendium of ERT Groundwater Sampling Procedures,* (EPA, 1991), *Compendium of Surface Water and Sediment Sampling Procedures,* (EPA, 1991), and *Compendium of ERT Waste Sampling Procedures* (EPA, 1991). Adapting sample collection procedures to accommodate chemical constituents of interest and sample type often may be required. This is acceptable as long as laboratory professionals are consulted and any deviations in procedure are documented, with rationales given for any changes.

Many considerations must be taken into account when developing sampling procedures and compiling equipment and supplies. Some of these considerations follow:

- Selecting and utilizing correct sampling procedures and establishing sampling plans that are written in a clear and easy-to-follow format.
- Using appropriate sampling devices and equipment, which in most cases should be disposable due to the impracticality of field decontamination between uses.
- Identifying proper sampling techniques for each type of site condition that may exist.
- Employing proper safety procedures and personal protective equipment (PPE) to ensure that personnel are not contaminated and exposure is minimized.
- Identifying potential sources of cross-contamination resulting from sample-handling errors such as placing a soil sample taken to determine trace levels of total petroleum hydrocarbons near a sample of pure fuel oil.

In later chapters, more detailed information is provided on the topic of quality assurance (QA) measures—those measures that must be employed to ensure that field-sampling errors are minimized and gross errors in procedure are identified and corrected. Measures that may be used include the following:

- Travel or trip blanks (uncontaminated material in identical sample containers, processed just as the actual samples).
- Equipment decontamination samples; that is, equipment rinseate sample, as required (whereby reusable equipment such as a hand-operated soil-sampling auger is washed with a decontamination solution, and then a sample is collected to determine the effectiveness of decontamination methods).
- Field documentation audits (periodic review of field notes, chain of custody records, sample collection or sample request forms, observation of sampling techniques, etc.).

FACTORS AFFECTING SAMPLE COLLECTION

Important sampling concepts are discussed in this section. First, there are several that have an impact on the sampling process and should be reviewed when attempting to determine how to obtain samples for analysis. These factors include the *amount* of material from which a sample is to be obtained, the *physical state* of the material, the *chemical properties* of the material, *environmental conditions* (i.e., temperature, wind conditions), and *concentration of contaminants or chemical constituents*, among others.

Amount of Material to Be Sampled

Many materials such as soils, shredded automotive parts, mining waste, and dredge spoils are stored in stockpiles. Obtaining representative samples of these materials is one of the sampling objectives. As an example, when samples must be obtained from large stockpiles of abrasive blasting media, you then must determine whether the sampled material represents the entire stockpile or whether portions of the pile will be considered batches from which samples are to be taken. This is especially essential in instances when one batch may be the result of a slightly different production process, thereby containing higher levels of certain contaminants. The entire stockpile also may be sampled to determine average or representative values for each of the contaminants. With proper sample collection and compositing (thoroughly mixing several separate samples) techniques, a representative sample may be obtained. Another good example of this type of sampling is the sampling of a large coal pile at a power station for the purpose of determining sulfur concentrations for the entire coal stockpile. A careful sampling technique involving timed-interval collection of material taken off conveyor belts, compositing, and then averaging laboratory-determined values from each composite group provides a representative value for sulfur content of the entire stockpile.

See Table 1–1 for preservation and other sample collection information.

Physical State of the Sample

The physical state of the material to be analyzed affects the techniques and equipment to be utilized in collecting the sample. Consider the effect that a two-phase (i.e., two substances of differing specific gravity and/or solubility) sample matrix

TABLE 1–1
Sample collection, preservation, and laboratory method chart.

Parameter	Solid Waste Method	Water/Waste-water Method	Sample Volume	Holding Time	Container Solid	Container Liquid	Pres. Liquid
Chromium hexavalent	7296	218.4	10g[a] 100mls[b]	24 hours	8 oz. CWM[a]	250 ml HDPE[b]	Cool 4°C
Mercury	7470	245.2	10g[a] 100mls[b]	28 days	8 oz. CWM[a]	250 ml HDPE[b]	$HNO_3<2$
Metals except Cr6+ and Hg	7000 series or 6010	200 series or 200.7	10g[a] 200[b] mls	6 months	8 oz. CWM[a]	500 ml HDPE[b]	$HNO_3<2$
Nitrate	9200	352.1 or 300.0	10g[a] 100mls[b]	48 hours	8 oz. CWM[a]	250 HDPE[b]	Cool 4°C
Nitrate/Nitrite	9200	353.3 or 300.0	10g[a] 100mls[b]	28 days	8 oz. CWM[a]	250 HDPE[b]	Cool 4°C $HNO_3<2$
Nitrite	—	354.1 or 300.0	10g[a] 50mls[b]	48 hours	—	125 ml HDPE[b]	Cool 4°C
Oil & grease/TPH	9070	413.1	10g[a] 1L[b]	28 days	2L CWM[a]	125 ml HDPE[b]	Cool 4°C $H_2SO_4<2$
Organic carbon	9060	415.1	10g[a] 25ml[b]	28 days	4 oz CWM[a]	125 ml HDPE[b]	Cool 4°C $HCl/H_2SO_4<2$
Orthophosphate	—	365.2/300.0	50 mls[b]	48 hours	—	125 ml HDPE[b]	Filter immed/cool 4°C
Sulfate	9035 to 38	375.4	10g[a] 50 mls[b]	28 days	4 oz CWM[a]	125 ml HDPE[b]	Cool 4°C
Sulfide	9030	376.2	100g[a] 500 mls[b]	7 days	8 oz CWM[a]	1 L HDPE[b]	ZnAc plus NaOHpH 9
Sulfite	—	377.1	50 mls[b]	Analyze immed	—	125 ml HDPE[b]	None
Surfactants	—	425.1	250 mls[b]	48 hours	—	500 ml HDPE[b]	Cool 4°C
Temperature	—	170.1	1 L[b]	Analyze immed	—	2 L HDPE[b]	None
Tox-Total organic halogen	9020	—	10g[a] 250 mls[b]	28 days	4 oz CWM[a]	16 oz B.R.[b]	$H_2SO_4<2$, Cool 4°C
TRPH	—	418.1 or 503E	1000mls[b]	28 days	—	A.J.	Cool 4°C HCLH<2

Parameter	Solid Waste Method	Water/Waste-water Method	Sample Volume	Holding Time	Container Solid	Container Liquid	Pres. Liquid
Turbidity	—	180.1	100mls[b]	48 hours	—	250 ml HDPE[b]	Cool 4°C
Acrolein and acrylonitrile	8030	603	5g[a] 40 mls[b]	14 days	4 oz CWM[a]	40 m Glass V[b]	.008%$NA_2S_2O_3$ pH 4.5
Aromatic hydrocarbons, polynuclear	8310	610	10g[a] 1L[b]	7ext./40 days aft.ex.	8 oz CWM[a]	1-1 LA.J.	.008%$NA_2S_2O_3$ dark
Aromatic hydrocarbons, purgeable	8020	602	5g[a] 40 mls[b]	14 days	4 oz CWM[a]	40 ml Glass V[b]	.008%$NA_2S_2O_3$ HCl<2
Benzidines	8270	605	10g[a] 1L[b]	7ext./40 days aft. ex.	8 oz CWM[a]	1-1 LA.J.	.008%$NA_2S_2O_3$ 4°C
Chlorinated herbicides	8150	509B[c]	10g[a] 1L[b]	7ext./40 days aft. ex.	8 oz CWM[a]	1-1 LA.J.	Cool 4°C
Chlorinated hydrocarbons	8120	612	10g[a] 1L[b]	7ext./40 days aft. ex.	8 oz CWM[a]	1-1 LA.J.	Cool 4°C
Dioxins and furans	8280, 8290	613, 1613	10g[a] 1L[b]	7ext./40 days aft. ex.	8 oz CWM[a]	1-1 LA.J.	.008%$NA_2S_2O_3$ 4°C
Halocarbons, purgeable	8010, 8240	601, 624	5g[a] 40 mls[b]	14 days	4 oz CWM[a]	40 ml Glass V[b]	.008%$NA_2S_2O_3$
Haloethers	8270	611	10g[a] 1L[b]	7ext./40 days aft. ex.	8 oz CWM[a]	1-1 LA.J.	.008%$NA_2S_2O_3$ 4°C
Nitroaromatics and isophorone	8090	609	10g[a] 1L[b]	7ext./40 days aft. ex.	8 oz CWM[a]	1-1 LA.J.	.008%$NA_2S_2O_3$ dark
Nitrosamines	8270	607	10g[a] 1L[b]	7ext./40 days aft. ex.	8 oz CWM[a]	1-1 LA.J.	.008%$NA_2S_2O_3$ dark
PCBs	8080	608	10g[a] 1L[b]	7ext./40 days aft. ex.	8 oz CWM[a]	1-1 LA.J.	Cool 4°C
Pesticides, chlorinated	8080, 8140	608	10g[a] 1L[b]	7ext./40 days aft. ex.	8 oz CWM[a]	2½ LA.J.[b]	Cool 4°C, pH 5–9

TABLE 1–1
(continued)

Parameter	Solid Waste Method	Water/Waste-water Method	Sample Volume	Holding Time	Container Solid	Container Liquid	Pres. Liquid
Phthalate esters	8060	606	10g[a] 1L[b]	7ext./40 days aft. ex.	8 oz CWM[a]	2-1/2 LA.J.[b]	Cool 4°C
Semivolatile	8250, 8270	625, 1625	10g[a] 1L[b]	7ext./40 days aft.ex.	16 oz CWM[a]	2-1/2 LA.J.[b]	Cool 4°C
TCLP 1311 extraction	—	100g[a]		7ext./40 days aft. ex.	2.5 L CWM[a]	—	None
TCLP sample	1311	—	100g[a]	7ext./40 days aft. ex.	16 oz CWM[a]	—	None
Volatile organics	8240	624, 1624	5g[a] 40mls[b]	14 days	4 oz CWM[a]	40 ml Glass V[b]	4°C
Volatile organic nonhalogenated	8015, 8240	—	5g[a] 40mls[b]	14 days	4 oz CWM[a]	40 ml Glass V[b]	4°C
Radiological test—gross alpha	9310, 9315	—	100g[a] 1L[b]	6 months	8 oz HDPE[a]	2 L HDPE[b]	HNO_3<2
Radiological test—beta	9310	—	100g[a] 1L[b]	6 months	8 oz HDPE[a]	2 L HDPE[b]	HNO_3<2
Radium (total)	9320	—	100g[a] 1L[b]	6 months	8 oz HDPE[a]	2 L HDPE[b]	HNO_3<2
Coliform-fecal and total	9131, 9132	909A, 909C[c]	n/a[a] 100mls[b]	6 hours	4 oz CWM[a]	250 ml HDPE[b] 4°C	.008%$NA_2S_2O_3$
Fecal Streptococci	—	910A, 910B[c]	–100mls[b]	6 hours	—	250 ml HDPE[b]	.008%$NA_2S_2O_3$ 4°C
Phenols	8040	604	10g[a] 1L[b]	7ext./40 days aft. ex.	8 oz CWM[a]	1-1 LA.J.	.008%$NA_2S_2O_3$ 4°C

[a] Solid waste method (SW-846)
[b] Water/Wastewater method (EPA-600)
[c] Standard methods, 16th ed.

Glass V = Glass VOA vial
HDPE = High-density polyethylene bottle
B. R. = Boston round

CWM = Clear wide mouth
A. J. = Amber jug/jar

may have on an analysis for mercury. For example, you must obtain a sample of a matrix consisting of water and a solid particulate material that has settled to the bottom of a tank. The particulate may contain substantially higher levels of mercury than the water phase of the matrix due to the attraction of mercury droplets to solid particles in preference to water. This necessitates obtaining a complete column of liquid that captures the particle settlement and submitting it for analysis to obtain a representative result.

Chemical Properties of the Material

The chemical properties of the material to be sampled also affect how samples should be collected. For example, it is desired that a sample of wastewater be obtained to determine the amount of volatile organic compounds (VOCs) present. By understanding that these materials readily evaporate, it is determined that a glass container with a Teflon-sealed cap called a *volatile organic analyte/analysis (VOA)* vial should be used. The objective of the sampling activity can be achieved due to the ability of the container to completely seal the volatiles in the sample matrix (*matrix* refers to the materials that make up the entire sample, such as soil, water/oil mixture). This sample container should be filled completely, forming a meniscus (convex surface) on top, and then sealed, thereby eliminating air space at the top of the vial into which organics may volatilize. A sample of wastewater contaminated with heavy metals involves other considerations, such as its pH. When sampling waters for heavy metals, they are collected in the appropriate containers and then acidified to a pH of <2 so that the metals in the water will remain in solution. If not acidified, the sample constituents may precipitate, "plate out," or adhere to the sides of the sample container. Another consideration is potential biodegradation of the chemical constituent (analyte) of concern, as in the case of samples of chlorinated water where trihalomethanes are the analyte. By cooling the sample, this process is substantially slowed or stopped. Other types of samples affected by chemical properties include materials subject to rapid oxidation, soil that may lose volatile constituents, biological samples, and water samples for pH and temperature.

Environmental Conditions

Conditions that exist at the location where samples are to be taken may affect the samples and should be documented on sample-collection logs and field notes. For example, higher than normal temperature may cause more volatilization of constituents in samples, resulting in lower than expected VOC values. Windy conditions could transfer contaminants out of areas with high contaminant levels and lead to cross-contamination of samples on which trace analysis is to be conducted for heavy metals.

Concentration of Constituent in a Sample

The concentrations of chemical constituents present in the various sample matrices also are worthy of consideration. Samples that contain high levels of contaminants pose possible handling, cross-contamination, and safety concerns. Care must be exercised when filling containers with these types of samples. If possible, do not spill or drip the sample onto the outside of the container because it could contaminate sample transport containers or laboratory work surfaces. If material is spilled onto the outside of the container, it should be cleaned and decontaminated after sealing as effectively as possible in the field, using appropriate solutions and clean wipes or towels. Solutions used for decontamination of sampling equipment, especially organic solvents such as hexane, iso-octane, and iso-propanol are potential sources of cross-contamination, for VOC or total petroleum hydrocarbon (TPH) samples. Due to this potential for sample contamination, many sampling technicians do not transport into the field any of these solvents. Many technicians and couriers do not even fuel their vehicles on days when they are collecting these types of samples, out of concern for cross-contamination. Remember that the laboratory often is capable of detecting parts per trillion.

In the event that the sample requires highly sensitive laboratory analysis, and thus is expected to contain little or only trace amounts of a contaminant, care must be exercised to avoid contact with items or areas where higher concentrations may be found. Cross-contamination may be caused by this contact or even through the sample technicians' gloves contacting areas that are highly contaminated. Samples such as wipe tests or liquid samples from COLIWASAs or thieves may easily be cross-contaminated due to inappropriate reuse of sampling devices or used gloves. Don a new pair of gloves for each wipe and use new drum thieves for each drum liquid sample. If using reusable sampling devices, proper decontamination techniques such as solvent rinsing, deionized water/Alconox, steam cleaning for reusable soil sampling devices (when sampling for organics), or acidified water (HNO_3 pH < 2) rinse for a reusable water-sampling device should be employed. Decontamination of equipment is discussed in more detail in the section on Care and Decontamination of Equipment in Chapter 2. The effect of cross-contamination due to improper sample handling and collection may be determined through the use of rinseate blanks, which are discussed in the section on quality assurance (QA) and quality control (QC) in Chapter 4.

REPRESENTATIVE SAMPLING

Samples of the entire quantity of material collected should be essentially similar in chemical and physical properties to ensure that they are "representative." *Representative sampling* refers to the collection of an amount of sample (aliquot) that possesses properties and concentrations similar to the entire sample matrix (i.e., body of water, waste pile). Another way of defining representative sample is that the sample is representative of the entire quantity of sampled material.

Heterogeneous materials are those that are dissimilar from location to location within the material. Examples of this include water and oil mixed in a barrel, and a barrel with solid and liquid that has separated due to gravity. To properly sample these types of materials, a representative column of the liquid and solid in the container should be obtained. Under certain well-controlled circumstances, the sample may be mixed or made into a similar state; that is, mixed or agitated (homogenized) prior to collecting the sample. Otherwise, agitation or mixing should be avoided since it will result in a sample that is not representative.

In the event that the material to be sampled exists as a homogeneous mixture, a representative sample easily can be taken. In a heterogeneous sample, if only one phase in an oily water mixture was sampled, then a partitioning effect may be observed in contaminant levels that are measured. *Partitioning* refers to the transfer of containments from one phase to another. For example, if polychlorinated biphenyls (PCBs) are being analyzed in the laboratory, it is likely that higher concentrations will be found in the oil phase that has separated from the water. This is due to the PCBs' relative insolubility in water and much higher solubility in oil. Care should be taken not to "stir" a heterogeneous mixture and then attempt to sample it. This will result only in all samples being different in nature. A material may be mixed or agitated only under carefully controlled conditions where it is thoroughly agitated with a device specifically designed for that task. Alternatively, cross-sectional sampling methods may be employed (i.e., COLIWASA for shallow samples of less than 48 inches or bacon bomb for depths of more than 48 inches).

Homogeneous materials do not need to be agitated or mixed prior to sampling because of the uniform consistency of the material throughout the mixture. Once verified as homogeneous, these materials may be sampled at any time. A single grab sample (*grab* refers to a sample taken at one location at a specific time) of a homogeneous material is considered representative.

SAMPLE SELECTION METHODS

The appropriate number and location of samples may be established by several methods. Due to variable site conditions and analytes, an experienced laboratory or environmental professional should be consulted to formulate sample collection strategies. Limitations exist with all of the methods. This is due to assumptions that have to be made about the sample population *(N)*. When using random sampling, we assume that contaminant distribution is uniform in the sample population. This is clearly not the case in the real world. Often contaminants or analytes are clustered or concentrated in limited areas (i.e., the needle in the haystack analogy). This complicates the task of designing a sampling plan that is defensible and reasonable in cost. For example, a leaking container of methylene chloride has been transported around an unpaved facility within 100 feet of a repair shop, thereby contaminating significant patches of soil. If samples were taken randomly throughout the entire yard, little contamination would be found because most of it exists in limited areas within 100 feet of the shop. See Figure 1–1.

In the most basic situation, such as a preliminary determination of the existence or presence of a pollutant at a discrete location, a judgment may be made about how and where the material was released or spilled (*judgmental sampling*). The basis of this method is applying judgments or estimates based on available information, such as the areal extent of contamination, visible traces, or known concentration of materials spilled or released.

The next type of sample selection is simple random sampling. *Random sampling* does not mean undisciplined or irregular sampling. An example of random sampling in practice would be a surface-soil sampling plan for a large area that is known to be contaminated. An imaginary grid may be superimposed on the site, producing many discrete potential sample locations. These are numbered sequentially. The goal is to ensure that any part of the area being sampled has an equal chance for inclusion as a sample point. A random number table then is consulted to determine a suitable number and distribution of samples for collection. Remember that the assumption is that there is some uniformity of analyte or contaminant distribution. Statistics may be employed to determine the number of samples based upon a confidence interval

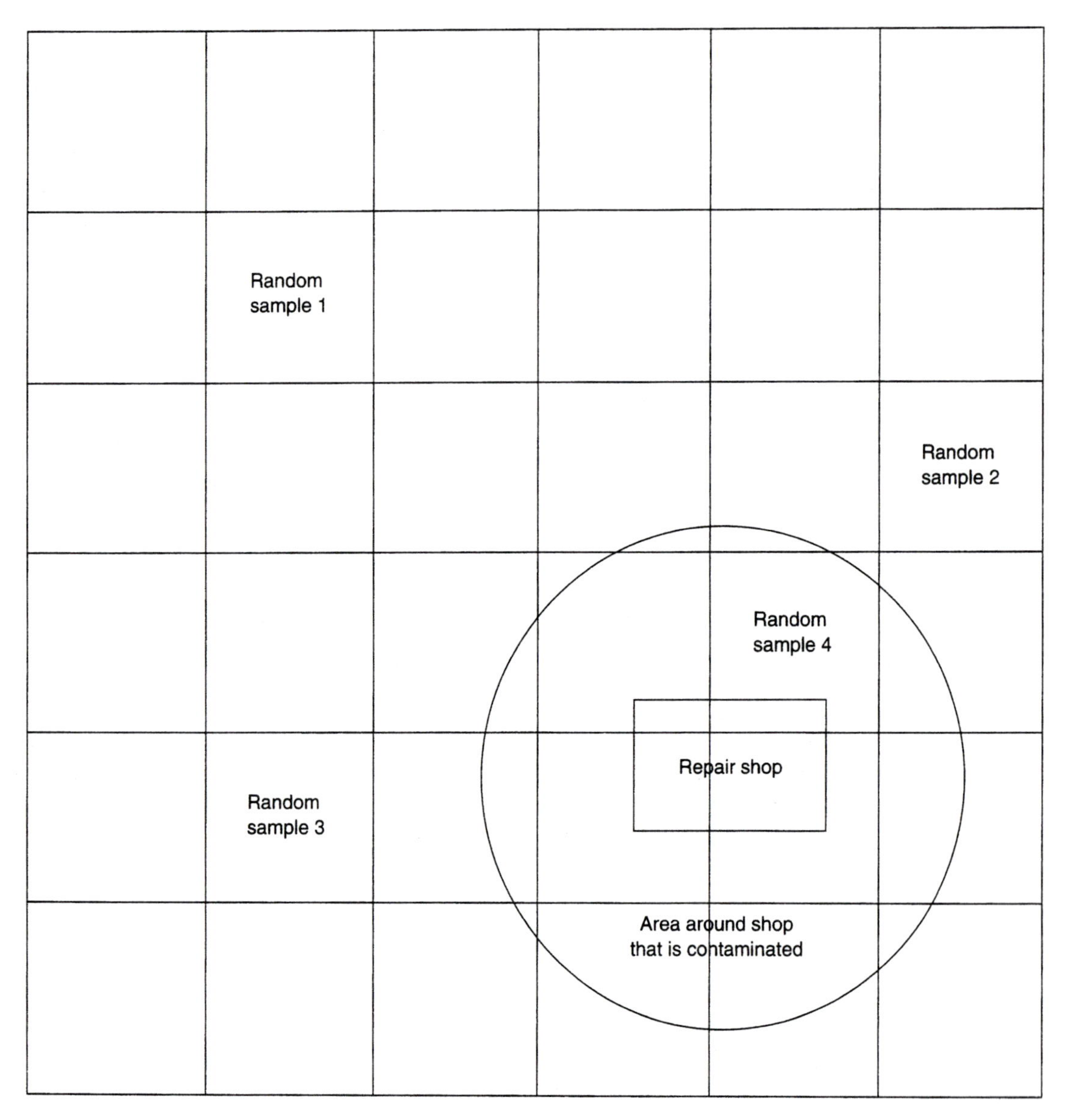

FIGURE 1–1
Contaminated areas around a repair shop that could be isolated for sampling, determined from such data gathering as interviews with employees, reviewing hazardous materials use patterns, and preliminary sampling.

(CI, which is a percentage of samples that will achieve adequate precision). A common misapplication of a statistical test requiring normally distributed data (e.g., a student's *t* test) occurs with many types of sampling where actual values demonstrate significant variation and occurrence within the sample population. A correct use of the student's *t* test is to determine the proper number of samples to be obtained from a source such as a wastewater discharge from a treatment plant. Actual values may vary above and below average in a pattern consistent with a normal distribution. In this case, the moderate variations above and below average values allow the student's t test to require a reasonable number of samples.

Another type of sample selection is *stratified random sampling.* In this method, samples are selected from every batch or strata of a material to be sampled to isolate the known variation between each. Simple random sampling then is applied within each batch or strata. As an example, ash residues have been sluiced (water added and then pumped) to an ash pond over many years, and it is known that chemical composition varied seasonally. That is, as coal was burned, the ash produced was significantly different in chemical composition from batch to batch or supplier to supplier. We may then identify the layers or strata in the ash pond and sample within each strata using random selection techniques.

Yet another type of sample selection is *systematic random sampling.* This method is employed by identifying a random sample point and then determining subsequent sample points by taking samples at fixed intervals from the original point. It also may be done by employing a formula or mathematical model to determine sample points. This may involve the use of square, rectangular, radial, or other grid configurations, depending on physical layout of the site. An example of this type of sampling is found in the *Field Manual for Grid Sampling of PCB Spill Sites to Verify Cleanup* (EPA, 1993). See Figure 1–2.

In the case of the EPA's PCB spill sites manual, the method for determining sample frequency and locations is based upon a grid size and pattern that is designed to locate spills 2 feet in diameter. Thus, it is apparent that in this method assumptions must be made about the contaminated sites. Such reasonable assumptions fit into a statistical model for locating contaminated areas to a certain degree of confidence. This method, when properly implemented and documented, is easier to defend than the other methods discussed here.

To determine the number of samples required to properly characterize the contaminants of a wastewater discharge or continuous waste generation process where several preliminary analysis results have been obtained, a statistical method involving the use of the students' t value and the variance in reported values can be used, as in the following example:

Given:

Ex = sum of measurement values

n = number of measurements

x = reported values for each measurement

$(Ex)^2$ = square of the sum of measurement values

Ex^2 = sum of squared measurement values

x = mean

RT = regulatory threshold

s^2 = variance

t = t value taken from students' t table (consult statistics text)

The following step-by-step procedure may be followed when establishing sample locations and determining numbers of samples for areas that have been contaminated with PCBs. An example of an 8 × 10-foot rectangular spill site has been used.

1. The site should be measured with a tape measure to obtain overall dimensions.
2. It may then be drawn to scale on graph paper.
3. The center and radius of the sampling circle are determined on a separate diagram. Draw a line representing the longest dimension, L_1. Find the midpoint P of L_1.
4. Draw a second line, L_2, perpendicular to L_1, through point P. Line L_2 must extend to the boundaries of the site.
5. Find the midpoint C of line L_2. Point C is the center of the sampling circle. (In this example, points P and C coincide; however, they might not coincide for many other types of configurations.)
6. Measure the distance from point C to either end of L_1, which is the sampling radius r. The distance, r, should be measured to the nearest 1/16 inch.
7. Scale the radius r up to actual size. In this example, radius r is 3¼ inches on a scale of 1 inch = 2 feet. The actual sampling radius = 6½ feet = 3¼ inches × 2 feet/inch.
8. The number of samples to be taken should be determined next. The number of samples to be taken depends on the length of the sampling radius, r.

Sampling Radius, r (feet)	Number of Samples
< 4	7
> 4 to 11	19
> 11	37

Since r is 6½ feet, the number of sample points is 19.

9. Plot the sampling points on the site diagram. The sampling points on each row of samples are a distance, s, apart. Each row of sampling points is a distance, u, apart. Use the following table to determine these distances.

Number of Samples	Distance, s, Between Points on a Row	Distance, u, Between Adjacent Rows
7	0.87 r	0.75 r
19	0.48 r	0.42 r
37	0.30 r	0.26 r

Note that adjacent sampling points in a row are staggered and that the sample points of one row are located midway (horizontally) between the sample points of the adjacent row.

10. Mark the sample points on the site starting at the center C, and then mark the middle row points a distance of 3 feet 2 inches apart. Locate the adjacent rows a distance of 2 feet 9 inches from the middle row and mark the four sample points in each of these rows a distance of 3 feet 2 inches apart. Note that not all 19 sample points fall within the sampling area. This is due to the fact that from each sampling point, an imaginary circle is drawn to represent a spill area. The circle may go outside the sampling site, but one sample would fall into this imaginary circle. Theoretically, this would capture a spill of such a diameter.

FIGURE 1–2
PCB (Polychlorinated biphenyl) sampling design.

As an example, you desire to calculate how many samples should be taken from a wastewater discharge. The regulatory threshold is 6.0 mg/L. Preliminary data have been obtained for the waste stream and indicate normally distributed values. Remember that one of the limitations of this method is that it should be applied only to data that are representative of a normal distribution or bell curve.

Values for Preliminary Samples	
3.1 mg/L	squared = 9.6
4.2 mg/L	squared = 17.6
6.5 mg/L	squared = 42.3
5.2 mg/L	squared = 27.0
4.8 mg/L	squared = 23.0
$Ex = 23.8$	$Ex^2 = 119.5$

Formula A

$$\text{Variance}(s^2) = [Ex^2 - (Ex)^2/n]/n - 1$$
$$\text{Variance} = [119.5 - (23.8)^2/n]/5 - 1$$
$$\text{Variance} = 1.56$$
$$n = 5$$
$$n - 1 = 4$$
$$x = 23.8/5 = 4.8$$
$$s^2 = 1.55$$
$$RT = 6.0$$
$$df = \text{degrees of freedom} = n - 1 = 4$$
$$t\ \text{value} = 2.132 \text{ at } 90\%\ \text{CI}$$

To calculate the proper number of samples that should be taken, use the following formula:

Formula B

$$\text{Number of samples } (N) = (t^2 s^2)/(RT - x)^2$$
$$N = [(2.132)^2 1.55]/(6.0 - 4.8)^2$$
$$N = 4.89$$

Note that the formula is very sensitive to changes in variance. The larger the variance or average variation between individual sample values, the more samples will be required. Increasing confidence levels increases the t value, thereby requiring the number of samples to increase. Additionally, more samples may be added to compensate for inaccuracies in the preliminary sampling data.

Some of the key concepts in sample selection include the following:

- Assumptions must be made about the sample population, and the interaction of these assumptions on sampling strategy should be understood.
- To apply statistical tests or methods to the sample selection, data must exist in the form of preliminary sampling results or a previously performed analysis on similar samples.

- Random sampling does not mean undisciplined or irregular sample collection.
- Precision or nearness to actual sample values is generally increased with increased sample numbers.
- Number of samples and sampling program costs rise proportionally with increases in confidence intervals.
- Many types or groups of environmental samples do not exhibit analyte concentrations of normal distribution (i.e., excessive variance), so applying statistical tests may be difficult.

SAMPLE INTEGRITY

Sample integrity is another important concept in sampling activities. To ensure that samples are received by the laboratory in good condition with labels intact, within appropriate time limits, and with the sample as close as possible to the condition in which it was collected, certain procedures must be followed. These procedures include a chain of custody records (*chain of custody* refers to a document indicating who had possession of the sample at all times—to be covered in Chapter 5 in the section on documentation), appropriate sample seals (normally used when legal or enforcement considerations are in play), sample preservation procedures (i.e., cooling to 4°C or adding preservation solutions), and using appropriate sample containers. These measures help to ensure that the quality and integrity of samples are maintained from the point of sampling to the analytical laboratory.

Analytical Laboratory Sample Management

A laboratory QA/QC plan is required of the commercial laboratory, as well as state certification for various types of samples. Make arrangements for laboratory acceptance of samples, including transportation. Determine packaging requirements based on analyses and laboratory protocol, and prepare chain of custody records for the samples. Preservation guidelines observed by the laboratory should be employed on all samples. Holding times also apply. In some cases, laboratory analysis must be performed within a short time frame—for instance, hexavalent chromium (Cr VI) must be transported to the laboratory within 24 hours or sample results may be invalidated. Samples of hazardous wastes are exempt from manifesting requirements when shipped to a laboratory or when being stored; however, storage time limits do apply to such samples.

Sample Container Labeling and Handling

Containers should be labeled and identified at the time of sample collection. Do not wait until returning to a staging area to identify samples. If they are not labeled and identified immediately, the chance of misidentifying the samples increases. Many technicians prelabel containers so that the chance of this occurring is minimized; however, care should be taken to ensure that the correct sample is placed in the prelabeled sample container. Sampled materials should not be allowed to contaminate the outside of the container, thereby possibly contaminating the laboratory or obliterating the label. Care also should be taken to minimize cross-contamination of certain types of samples. This may be prevented by decontaminating sampling devices, donning new gloves when handling each sample, and using disposable sample-collection devices. Contamination may come from sampling devices, sampler handling, or other samples or items in the immediate area of collection. This contamination may be airborne, such as a mist or dust. To avoid this, take samples to the staging area as soon as possible after collection, place them in the appropriate transport containers, and ready them for transportation.

UNITS OF MEASURE

Measurements represent a basic component of sampling and analysis and re-lated chemical concepts. These measurements allow us to express what we are attempting to identify within the samples and chemical mixtures. Measurements that may be obtained from analytical laboratories and, in some cases, from field instruments can be highly precise due to the advances of such instrumentation. Extremely small quantities of PCBs may be detected. Units as small as micrograms may be detected on solid surfaces. Parts per million (ppm) and parts per billion (ppb) are typical units of concentration used in analytical chemistry. These units are based on volume-to-volume measurements; that is, for 1 ppm, the volume equivalent is 1 cubic centimeter/1,000,000 cubic centimeters (cm^3). Milligrams per liter (mg/L), a mass-per-unit volume measurement, is another common unit used in reporting laboratory results. However, detection of quantities this miniscule require exacting and careful attention to procedures and avoiding any sources of cross-contamination. Within this section, basic units of measure are reviewed and outlined.

Metric Units

Several systems of measurement are utilized in chemistry and sampling and analysis procedures. The metric, or SI, system is the primary set of units. This system is based on multiples of 10 and base units, which are considered standards. This is notably different from the English units that have arbitrary divisions and subdivisions, such as 1 pound, which is divided into sixteen parts called *ounces*. The metric system was recommended to the scientific community by the General Conference of Weights and Measures in 1960 to simplify the wide array of units that were in use at the time in engineering and science.

Base Units and Prefixes

A simple means of expressing very large or very small quantities, volumes, or distances is to use prefixes that have the meaning of times ten (× 10), times one thousand (× 1,000), or times one million (× 1,000,000), for example. Some of the basic units and prefixes are listed in Figure 1–3.

Table 1–2 represents a typical use of these concentration units when representing regulatory limits.

Base Units	Name	Abbreviation
Mass	Gram	g
Length	Meter	m
Time	Second	s (sec)
Temperature	Celsius	C (0°C = 32°F)
Volume	Liter	L

Prefixes of Units	Standard Abbreviation	Numerical	Text	Exponential
Mega	M	1,000,000.	million	10^6
Kilo	k	1000.0	thousand	10^3
Deci	d	.1	1 tenth	10^{-1}
Centi	c	.01	1 hundredth	10^{-2}
Milli	m	.001	1 thousandth	10^{-3}
Micro	μ	.000001	1 millionth	10^{-6}
Nano	n	.000000001	1 billionth	10^{-9}

Examples of Common Metric Units of Concentration

mg/m^3 (milligrams per cubic meter) used for gas vapor (such as mercury) in air, lead dust in air, etc. Example: 4.2 **mg/m^3.**

mg/kg (milligrams per kilogram) used for solids (such as lead in soil). This is also known as a ppm, or parts per million. Example: 55.3 **mg/kg.**

micrograms/cm^2 (**μg/cm^2**) used for the measure of a quantity of a substance present on a specific surface area. Example: 22.5 **μg/cm^2.**

mg/L (milligrams per liter) used for contaminants such as pesticides, organics, and heavy metals in water or other liquids. Example: 0.05 **mg/L.**

f/cc (fibers per cubic centimeter) used for asbestos and other fibers. Example: 300 **f/cc.**

FIGURE 1–3
Basic units of concentration, their prefixes, and common units.

TABLE 1–2
Inorganic persistent and bioaccumulative toxic substances.

	Soluble Threshold Limit Concentration Waste Extraction Test			Total Threshold Limit Concentration		
Analyte	Maximum Limit (mg/L)	Detection Limit (mg/L)	Analysis Result (mg/L)	Maximum Limit (mg/kg)	Detection Limit (mg/kg)	Analysis Result (mg/kg)
Antimony	15.00	0.10000	—	500	5.000	N.D.
Arsenic	5.00	0.10000	—	500	5.000	N.D.
Barium	**100.00**	**0.10000**	—	**10,000**	**5.000**	**16.00**
Beryllium	0.75	0.01000	—	75	0.500	N.D.
Cadmium	1.00	0.01000	—	100	0.500	N.D.
Chromium (VI)	5.00	0.00500	—	500	0.050	N.D.
Chromium	5.00	0.01000	—	2,500	0.500	N.D.
Cobalt	80.00	0.05000	—	8,000	2.500	N.D.
Copper	**25.00**	**0.01000**	—	**2,500**	**0.500**	**0.99**
Lead	5.00	0.10000	—	1,000	5.000	N.D.
Mercury	0.20	0.00020	—	20	0.013	—
Molybdenum	350.00	0.05000	—	3,500	2.500	N.D.
Nickel	20.00	0.05000	—	2,000	2.500	N.D.
Selenium	1.00	0.10000	—	100	5.000	N.D.
Silver	5.00	0.01000	—	500	0.500	N.D.
Thallium	**7.00**	**0.10000**	—	**700**	**5.000**	**25.00**
Vanadium	24.00	0.05000	—	2,400	2.500	N.D.
Zinc	**250.00**	**0.1000**	—	**5,000**	**0.500**	**14.00**
Asbestos	—	10.00000	—	10,000	100.000	—
Fluoride	180.00	0.10000	—	18,000	1.000	—

TTLC results are reported as mg/kg of wet weight. Asbestos results are reported as fibers/g. Analytes reported as N.D. were not present above the stated limit of detection.

QUESTIONS FOR REVIEW

1. What publications are available on the subject of sample collection, preservation, and analysis methods? List and describe three.
2. What does the term *two-phase sample* mean? Describe the ways in which a two-phase sample may present problems when sampling.
3. What is the chemical property of "volatility" and what type of sample may particularly be affected by it?
4. Why are preservation techniques used? Give examples of different types of sample preservation.
5. Cross-contamination is a significant potential problem when sampling. What types of things might lead to cross-contamination?
6. What is the process of decontamination. Give an example of a specific type of decontamination.
7. Representative sampling ensures that samples are__________?
8. How may a heterogeneous sample be described? Give an example of a heterogeneous sample.
9. What are the basic types of sample selection methods? What are the limitations of statistical methods?
10. What are the number of samples required for the following data set: 43 mg/L, 55, 65, 28, 51, 53, 62. Student's t at the 95% confidence interval. $t = 2.447$; $RT = 60$

ACTIVITIES

1. Contact a local environmental laboratory and obtain sample collection protocol for several types of samples that the laboratory is certified to perform. Prepare a brief report summarizing five types of samples (sample collection) and indicating what types of contaminants the lab may analyze for.
2. Obtain from a local environmental laboratory a list of analyses and detection limits with units of measure that are used to report results of analyses. Compare the list with an analytical report of the drinking water in your location. You may obtain the report by contacting the supplies of your drinking water.

READINGS

Compendium of ERT Groundwater Sampling Procedures, 1991. Environmental Protection Agency.

Compendium of ERT Waste Sampling Procedures, 1991. Environmental Protection Agency.

Compendium of Surface Water and Sediment Sampling Procedures, 1991. Environmental Protection Agency.

Compendium of ERT Waste Sampling Procedures, 1991. Environmental Protection Agency.

Standard Methods for the Examination of Water and Wastewater, 1992. 18th ed. Environmental Protection Agency.

Test Methods for Evaluating Solid Waste (SW-846), 1993. Environmental Protection Agency National Technical Information Service.

2

Sampling Equipment, Devices, and Containers

Patrick K. Holley

Upon completion of this chapter, you will be able to do the following:

- Recognize and describe various materials of construction of sampling devices and sample containers.
- Recognize, describe, and operate various sampling devices.
- Understand and apply the proper decontamination procedures to reusable equipment.

CONSTRUCTION MATERIALS AND COMPATIBILITY

Equipment and supplies must be carefully selected to properly collect and contain samples. The volume and type of sample to be taken, type of matrix, compatibility, and access to sample locations should be considered. Also, when selecting sampling equipment, chemical compatibility should be considered. Materials such as glass, Teflon, polyvinyl chloride plastic (PVC), polypropylene plastic (PPE), polyethylene (PE), polystyrene, brass, and stainless steel are often available. Glass is compatible with most materials (with the exception of hydrogen fluoride gas or solution). If glass containers are to be used for trace metals analysis, they should be made of borosilicate glass to avoid possible lead contamination. Polyethylene and polypropylene are available in a wide range of sample devices and are resistant even to organic solvents; however, over time they may swell or lose their dimensions and properties. Polyethylene is not to be used for trace level organic analysis due to the possible leaching of organics and plasticizers into the sample from the sampling device or the sample bottle. Additionally, organic analytes may be absorbed into the walls of the sample container. Polyvinyl chloride may be degraded by chlorinated solvents or aromatic solvents. Teflon is highly resistant to most chemicals; however, it is expensive and is somewhat soft or flowable, with a tendency to lose its shape. Stainless steel is used to build reusable devices for many sampling activities. The primary caution when using stainless steel is to avoid hydrochloric acid and chloride solutions, due to the pitting and corrosion problems they cause.

Sample quantity is determined by both the laboratory and the analytical procedure that has been selected for any given sample. In general, water samples require 1 liter sample quantities, except for volatile organic analysis, which uses a special 40-milliliter vial (VOA vial). Most individual soil analyses require less than 50 grams in the analytical laboratory; however, to allow for multiple analysis, quality control, and replicates, 200 to 1,000 grams of sample are collected. Oil samples generally require very small amounts to run analytical tests, with a sample volume usually of 40 to 50 milliliters. Consult the laboratory prior to sampling for exact volumes needed (the chart in the section on preservation and handling also may be used). Additionally, method manuals such as *Test Methods for Evaluating Solid Waste—SW-846* (EPA, 1993) list required specifications for sample quantities.

Sample containers must be properly cleaned prior to use in accordance with EPA protocol. The laboratory that will process and analyze the samples also should supply the sample containers. This way, lab protocol is more likely to be followed for proper container preparation.

MANUAL OR HAND-OPERATED EQUIPMENT AND DEVICES

Many devices are available for collecting samples. These devices should be selected based on their compatibility with the sample matrix, type of sample being collected, and field practicality. When conducting sampling activities, you may discover that certain tools, devices, and equipment are not adequate for proper collection, in which case an alternative device should be selected.

Glass Thief

A *glass thief* may be used to take a cross-sectional sample by lowering it slowly to the bottom of a container while ensuring that the tip of the thief does not touch the container's sides. Once fully lowered into the container, seal the upper end of the tube with your thumb (be sure you are wearing a glove). Then carefully withdraw the tube and place it into the sample container. Release your thumb and allow the liquid to flow into the sample container. Due to the negative pressure created inside the upper part of the tube, which does not contain liquid, the sampled liquid is held inside the tube; however, some sample liquid may run out of the bottom of the tube more readily with increasing columns of liquid or larger bore tubes (a stopper may be used carefully—too much force will break the tube—to seal the top of the tube if the thumb method does not suffice). Therefore, deeper containers (more than 36 inches) are not readily sampled with this method. Figure 2–1 demonstrates the step-by-step procedure for using a glass thief to collect a cross-sectional sample.

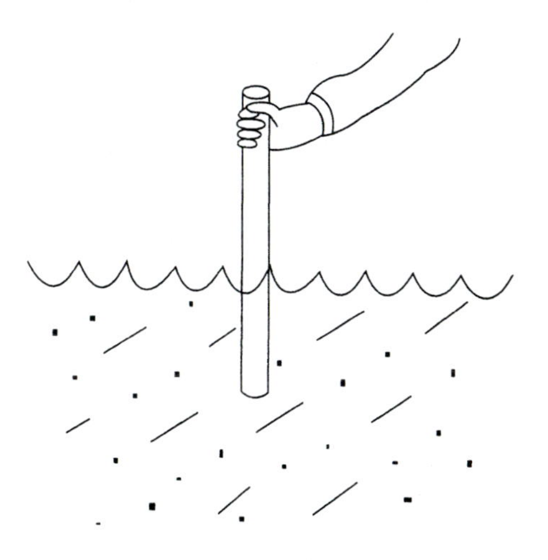

1. Insert open tube (thief) sampler in containerized liquid.

2. Cover top of sampler with gloved thumb.

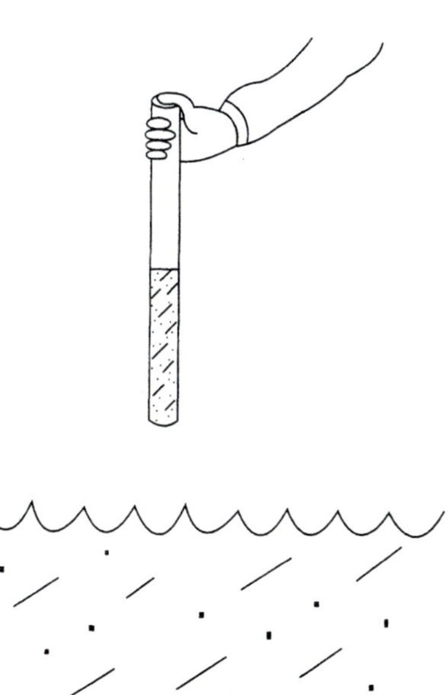

3. Remove open tube (thief) sampler from containerized liquid.

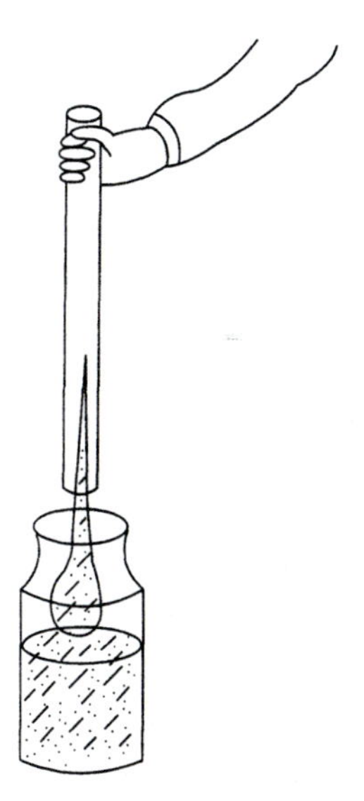

4. Place open tube sampler over appropriate sample bottle and remove gloved thumb.

FIGURE 2–1
The step-by-step procedure demonstrates using a glass thief when collecting a cross-sectional sample.

COLIWASA

A preferred method for drums and larger containers is the use of a *composite liquid waste sampler* (COLIWASA). The COLIWASA was devised to capture a representative column of liquid from various sized containers. These devices are made from a variety of materials—glass, polyethylene, Teflon, and others—and are equipped with a closure valve or ball at the tip of the tube. Once lowered into the liquid, the ball or valve, which is connected to the top of the COLIWASA via a rod, is closed, thereby sealing liquids inside the tube for withdrawal from the container. Prices for these devices range from $15 (for disposable units) to several hundred dollars, depending on size and construction materials. Reusable COLIWASAs must be thoroughly decontaminated between uses. The cost effectiveness of decontaminating these devices should be evaluated. Often the labor cost and waste disposal cost of decontamination solutions may exceed the cost of disposable units. See Figure 2–2.

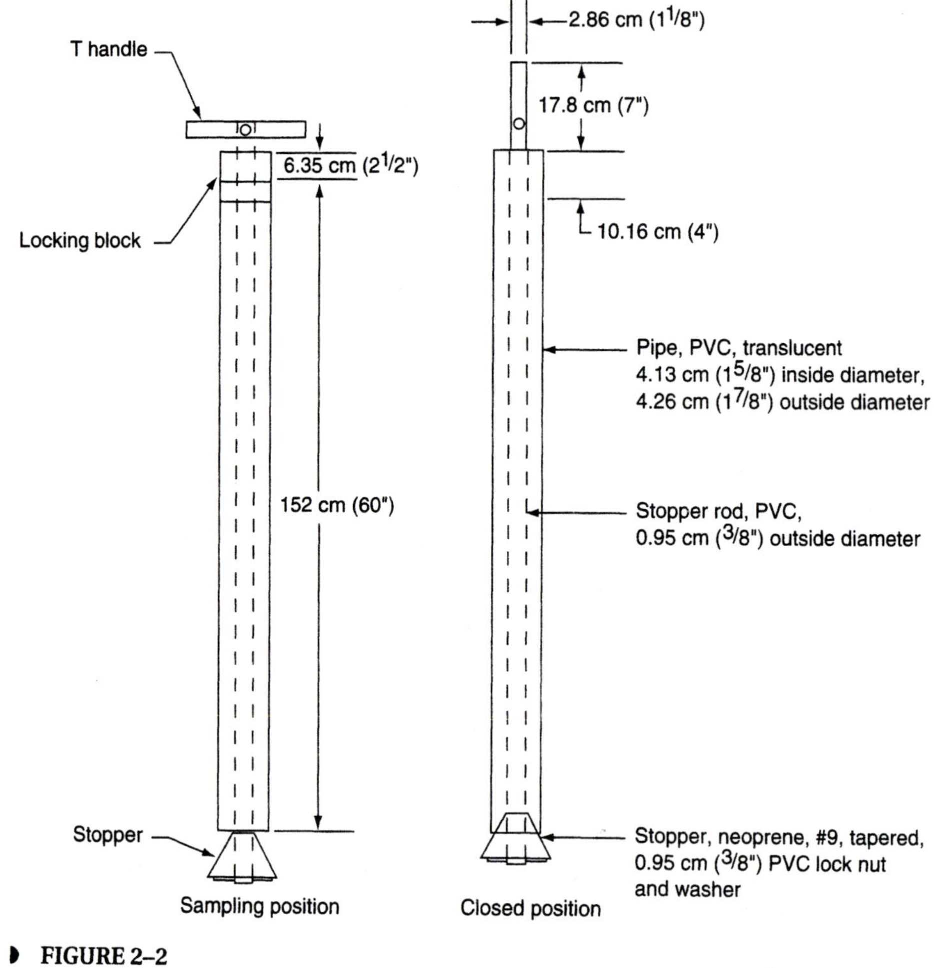

FIGURE 2–2
COLIWASA.

Triers and Grain Samplers

Granular or finely divided materials may be sampled easily with triers (similar to a small garden spade), grain samplers, or trowels. A trowel is simply a small hand-held tool for scooping soils or other solid materials. Triers are good for soft soils where samples will be taken near the surface. The trier consists of a tube that has been cut in half lengthwise, with a sharpened tip that allows the sampler to cut into sticky solids and to loosen hard soils. The trier is inserted into the material and the handle is twisted, thereby capturing a core. *Grain samplers* are for loose, flowable material such as sand, or for other materials that flow readily but that must be sampled deeper than surface level. See Figure 2–3. A grain sampler consists of a hollow rod with cutouts at about 6-inch intervals so that when the sampler is inserted into the pile and twisted, a uniform sample is captured inside the tube. See Figure 2–4.

Soil Augers or Tube Samplers

The *auger* consists of sharpened spiral blades that are attached to a hard metal central shaft; the blades cut into the soil when manually rotated using a T handle. Augers are not to be used for volatile organic samples because as they cut into the soil, they loosen the material, which may allow volatile components to evaporate. *Tube samplers* (or *thin-walled tube samplers*) are used for soils and solids, especially when volatile organic constituents are part of the analysis. These samplers can cut smooth cores from surface level to about 5 feet in depth. See Figure 2–5.

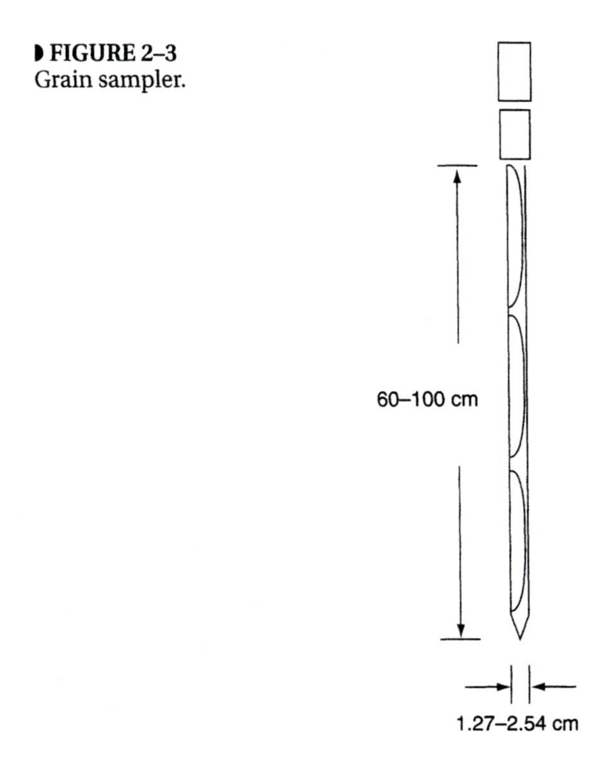

FIGURE 2–3
Grain sampler.

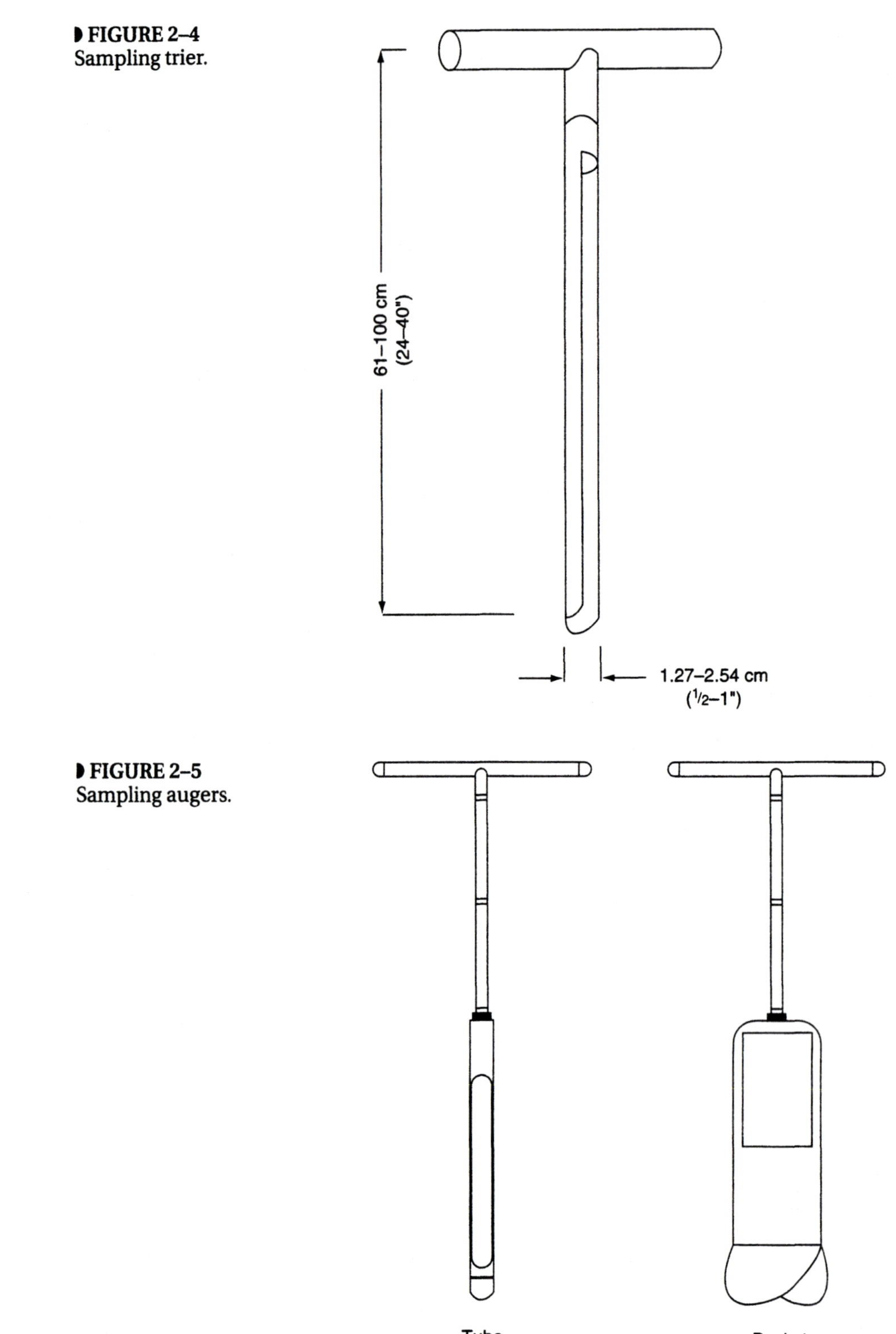

FIGURE 2–4
Sampling trier.

FIGURE 2–5
Sampling augers.

Weighted Bottle

A *weighted bottle sampler* is designed to collect samples of liquid from specific depths. The device is a bottle with a surrounding metal cage or containment that allows the operator to sink the sampler to a certain depth in a body of water and then pull a line connected to a stopper that allows water to flow into the bottle. See Figure 2–6.

Bacon Bomb

A *bacon bomb sampler* is similar to the weighted bottle except the device is constructed entirely of metal and has two built-in valves with a line connected to them via a rod. As the device is lowered to the appropriate depth, a line is pulled, opening the valves. This allows a uniform liquid sample to flow into the sampler as the air that was carried down by the sampler is released through the top valve. See Figure 2–7.

Dipper

A *dipper* is a simple device that is used to collect samples from a uniform or homogeneous liquid body. It consists of a long rod connected to a cup that will hold approximately 250 to 500 milliliters. Extensions may be included with the dipper for reaching laterally or downward to basins, tanks, sumps, or other containments from which samples are to be taken. This simple device often is used in wastewater treatment plants where bodies of water are uniformly agitated or continuously mixed. See Figure 2–8.

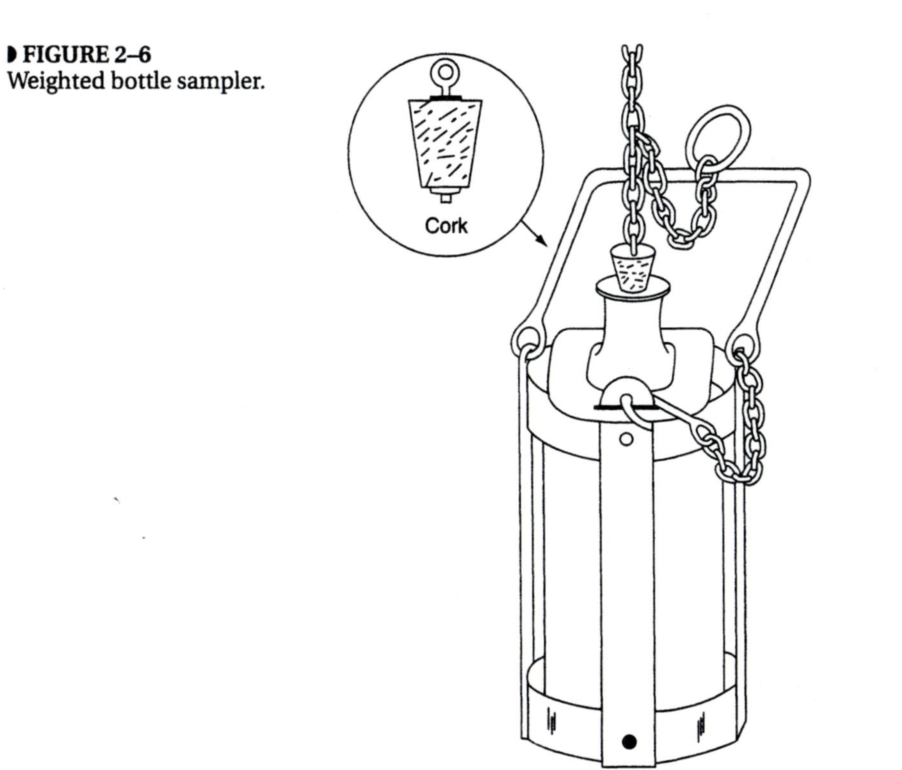

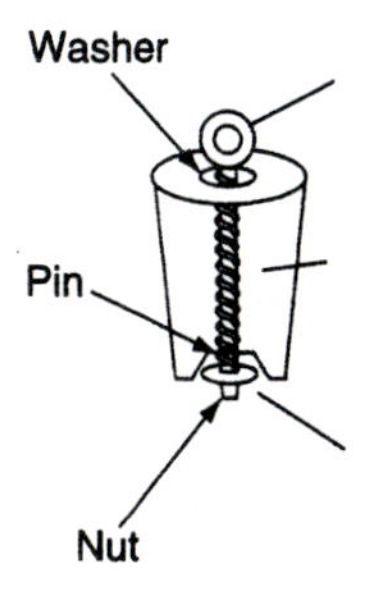

FIGURE 2–6
Weighted bottle sampler.

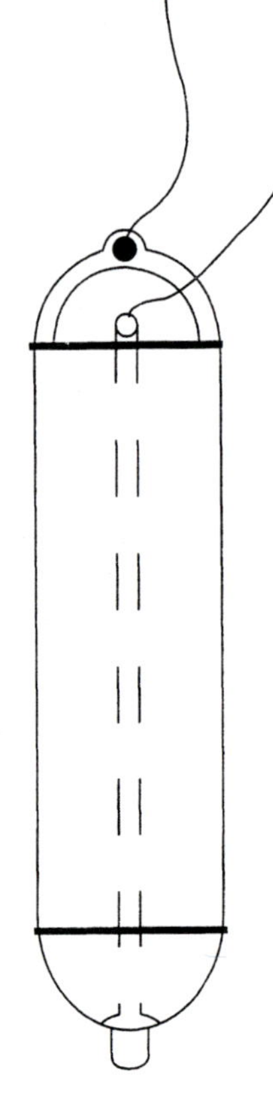

FIGURE 2–7
Bacon bomb sampler.
A bacon bomb is composed of metal, with two built-in valves. When the bomb is lowered to the right depth, a line connected to the valves via a rod is pulled, opening the valves and allowing the liquid sample to flow in.

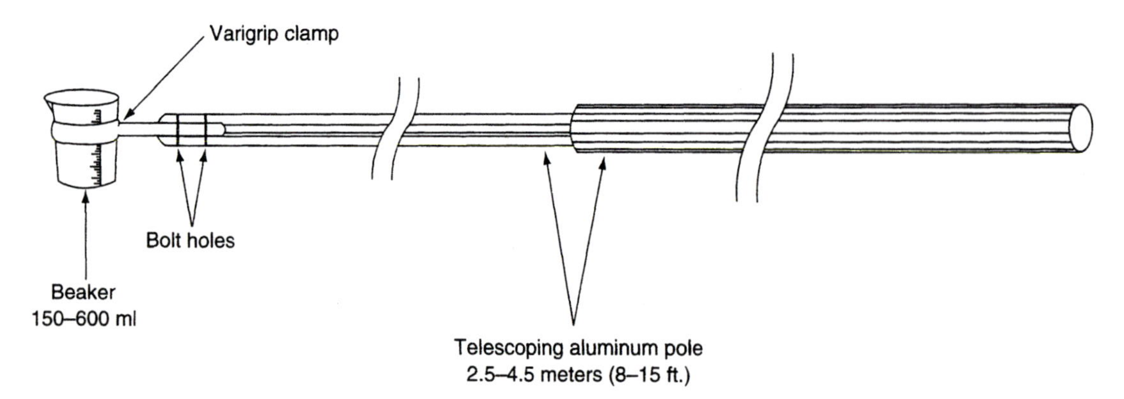

FIGURE 2–8
Dipper.

Remote Sealing Grab Sampler

A *remote sealing grab sampler* consists of a glass sample container connected to a polyethylene or Teflon body and a screw-on capping rod operated by the person collecting the sample. The bottle is lowered to the desired level, allowed to fill, and the cap is screwed in place and the device is raised to the surface. Both the bottle and cap should be compatible with materials that may be encountered in the sample matrix. See Figure 2–9.

Bailers

A *bailer* consists of a tube or cylinder that has a check valve in the bottom of the unit. It also has an open top with an attached rope or cable for lowering and raising the device into a groundwater well. When water contacts the bottom of the sampler, the check valve opens, allowing the water to flow up into the sampler. When lifted up, the sampler check valve closes and does not allow water to escape. To avoid cross-contamination on subsequent samples, the device must be decontaminated. The rope or cable also must be discarded or decontaminated for reuse. Boilers are manufactured in many configurations and of many different materials. Figure 2–10 is an example of a typical bailer used to sample groundwater wells.

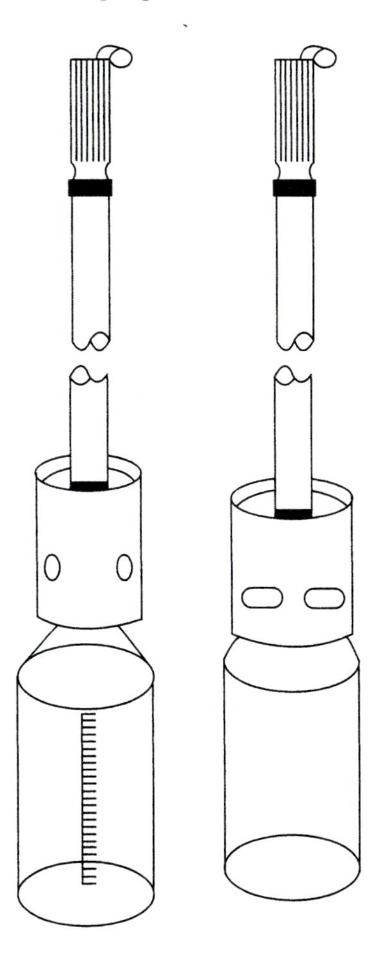

FIGURE 2–9
Remote sealing grab sampler.

FIGURE 2–10
Bailer.

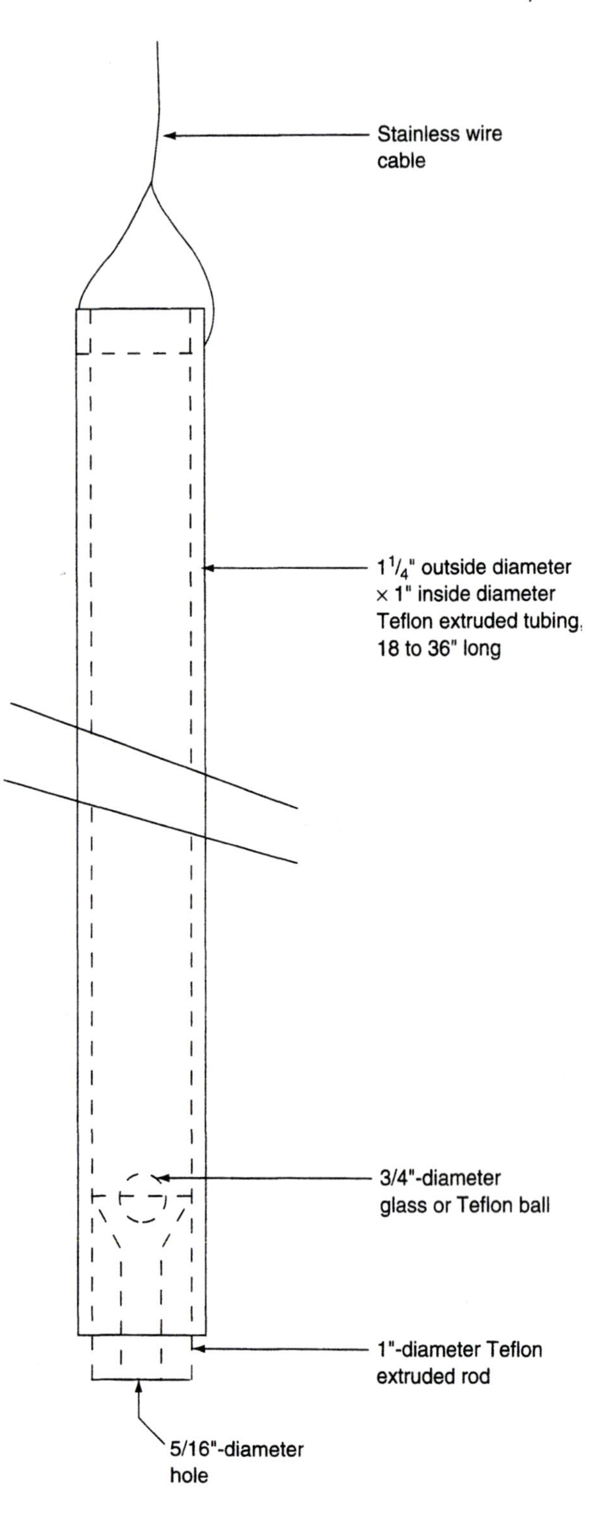

Ponar Dredge

A *ponar dredge* is used to collect upper-level sediment or sludge samples from water bodies such as tanks and ponds. The device resembles a clamshell bucket used in dredging harbors and channels. It is equipped with a spring-loaded tensioner that forces the jaws of the device to close once it has been lowered into the water and dropped sharply into the sediment. As the device hits the bottom, the spring tension is released, allowing the jaws to shut tightly and collect a sample. The device is lowered and retrieved by means of a cable or nylon rope. See Figure 2–11.

Coring Device

A *coring device* is used to collect samples from beneath the uppermost layers of sediment or sludge. It is similar to the thin-walled tube sampler used in collecting soil samples. The body of the device is attached to a handle or extension, and then a sleeve is inserted into the body and capped with a coring point. This coring point is screwed on to hold the sample sleeve in place. In cases when sediment conditions

FIGURE 2–11
Ponar dredge.

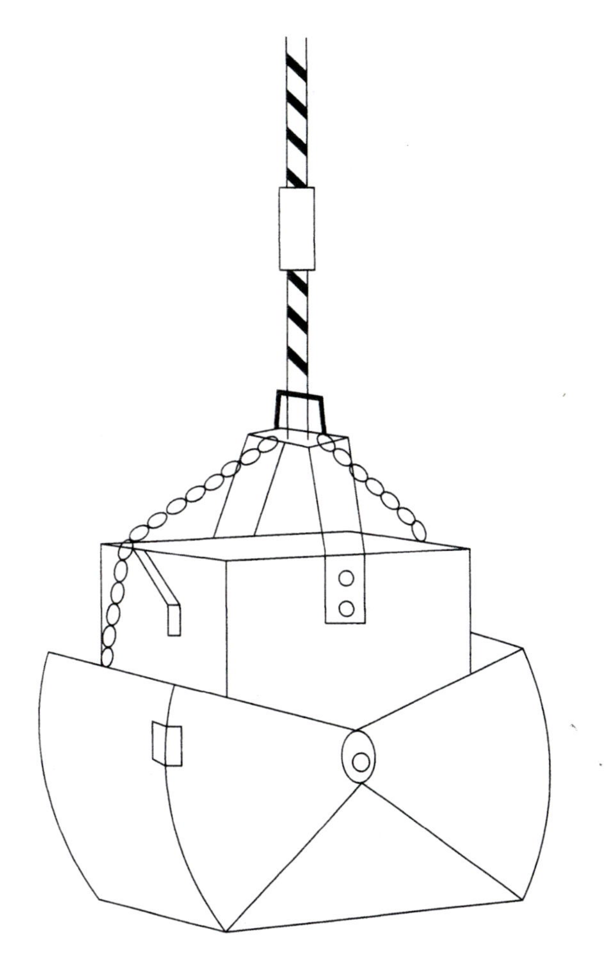

are very soft or difficult to contain in the sample sleeve, an "eggshell" insert is placed into the tip just behind the coring point. The flexible fingers of the eggshell insert allow the sediment or sludge to enter the tube easily and then hold the material in the tube when raised to the surface. See Figure 2–12.

POWER-DRIVEN SOIL SAMPLING EQUIPMENT

In situations when large numbers of samples or sample matrices are physically difficult to penetrate, *power-driven equipment* is often used. This equipment ranges from the traditional drilling or boring rig to electrically powered groundwater and well

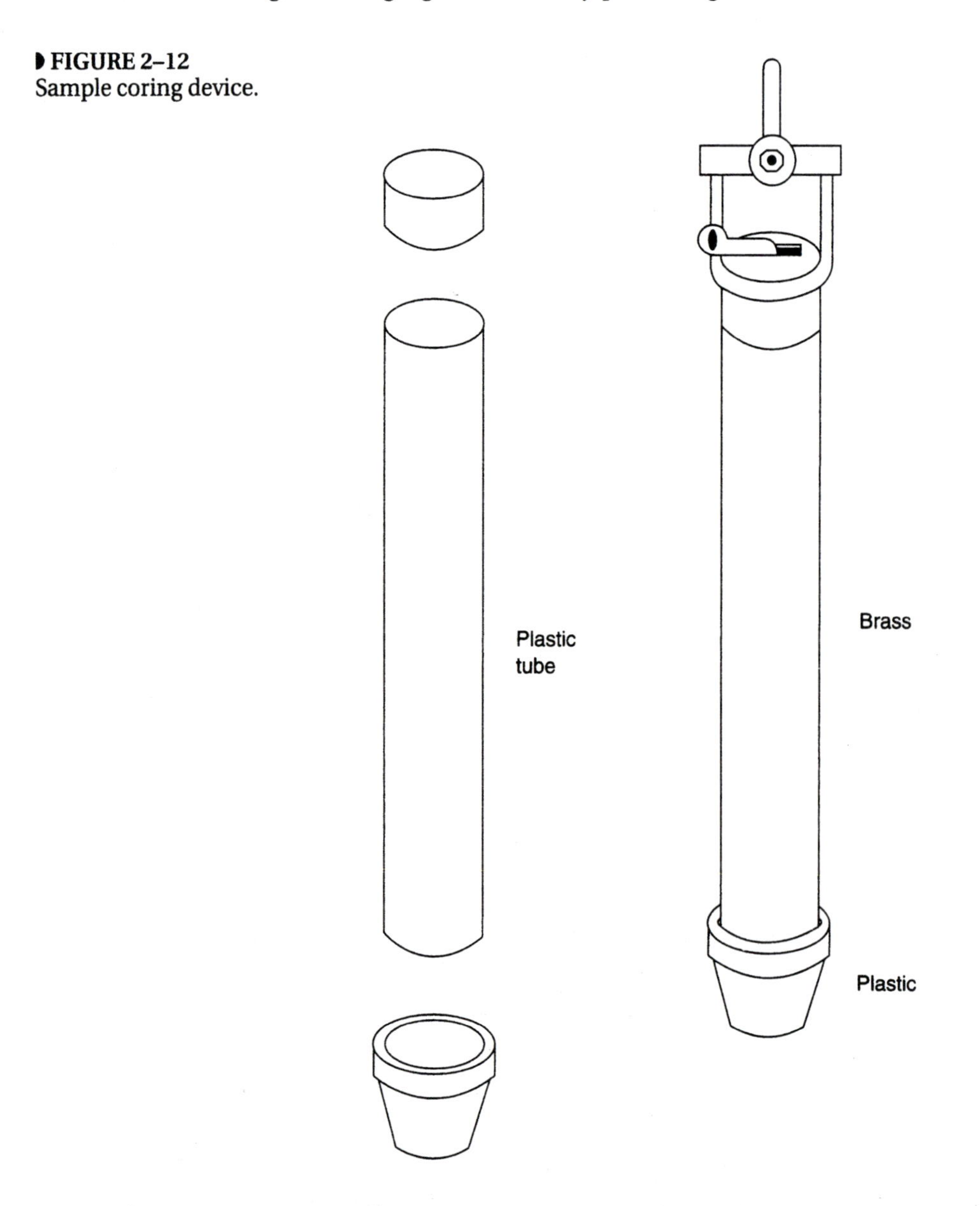

FIGURE 2–12
Sample coring device.

water sampling pumps. This equipment often adds speed, convenience, and reproducibility to a sampling program. See Figure 2–13.

Solid Stem Auger with Split Spoon Sampler

Solid stem augers are power-driven boring devices that normally are used to bore to the appropriate sample depth, and then a tube sampler or a split spoon sampler is inserted into the borehole to obtain the sample. The *split spoon sampler* is composed of two vertically split halves of a cylinder 2 inches in diameter and 12 to 18 inches long that are held together with threaded couplings installed over the top and bottom, or cutting end. Extension drilling rods are affixed to the top, and a slide hammer is used to drive the split spoon sampler into the soil at the selected depth. A check valve at the top of the sampler allows air to escape while the sampler is pushed into the soil. Once the sample is taken, the split spoon sampler is taken apart by removing the threaded ends and separating the cylinder halves. See Figure 2–14.

Hydraulic-Powered Impact Subsurface Samplers

Other power equipment, which goes by various trade names, is also available. One example is a truck-mounted tube sampler that is capable of being deployed to remote job sites. This sampler (brand names include Geoprobe™ and Earthprobe)

FIGURE 2–13
Power-driven soil sampling equipment. *Photo courtesy of the U.S. Navy.*

FIGURE 2–14
Split spoon sampler.

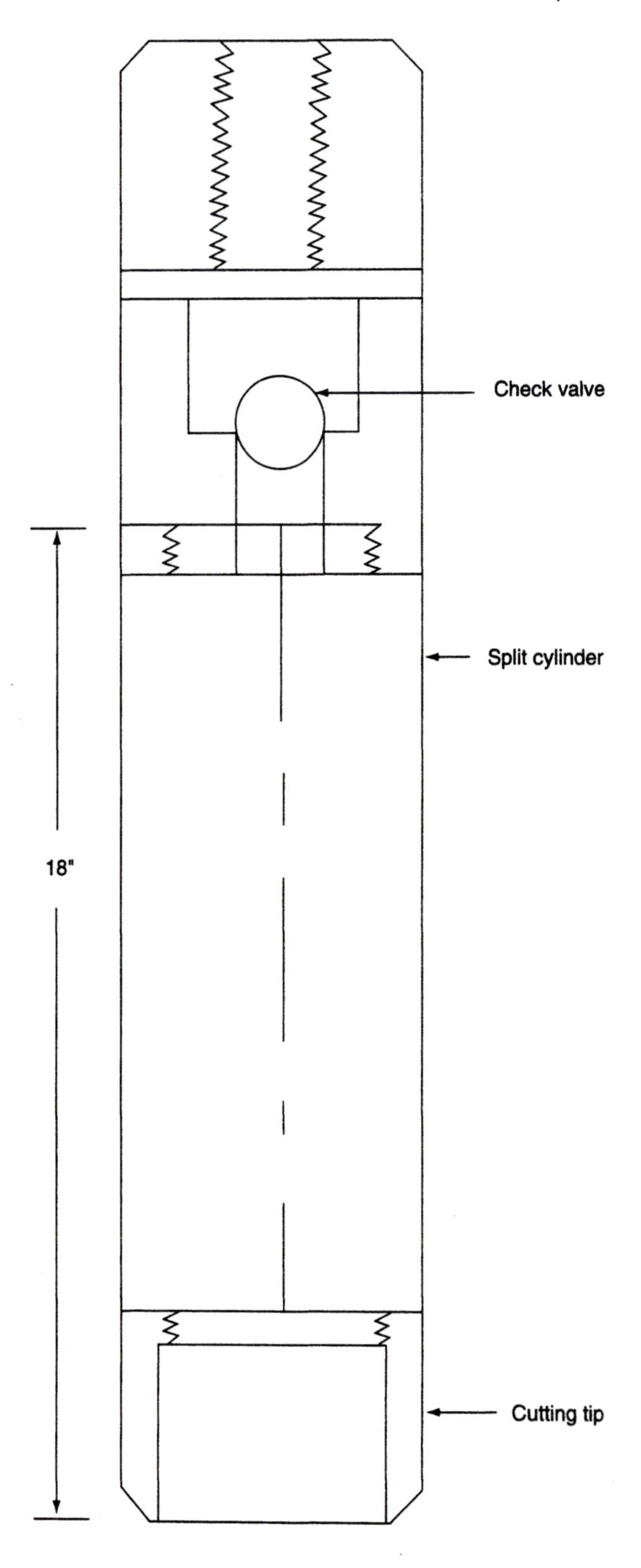

consists of a hydraulic-powered tilt-up boom and a hydraulic-impact hammer that drives a tube-type sample probe into the soil. These types of devices are capable of producing large numbers of high-quality samples in very short periods of time. See Figure 2–13.

Bladder Pumps

Bladder pumps are commonly available downhole devices used to sample wells. They are reliable and simple, and are operated by pushing compressed air into a chamber containing a bladder that fills with well water. As the air is pushed into the chamber, it pushes the well water out of the bladder and to the surface through a Teflon hose attachment. As with all other sampling devices discussed in prior sections, bladder pumps must be selected according to compatibility with the type of analysis being conducted. Increasingly, facilities involved in extensive groundwater monitoring programs are utilizing dedicated pumps (i.e., a pump is left installed in each well).

Centrifugal Pumps

Centrifugal pumps are devices that are positioned down the well casing where they pump liquid from the water in the well back to the surface at a continuous rate. A centrifugal pump with an impeller pushes water through a tube to the surface for collection. These devices can be used on very deep wells and provide a continuous flow of water over long periods of time.

CARE AND DECONTAMINATION OF EQUIPMENT

Sample devices may be decontaminated in the field using water-based solutions such as soap (Alconox or equivalent) and repeated deionized water rinses. Those contaminants that are caked or tightly adhered to the surfaces of the sampling devices may be water blasted or steam cleaned and then washed or rinsed in deionized water. A 10% nitric acid rinse followed by repeated, thorough deionized water rinses should be performed on stainless steel sampling devices that may have been contaminated with metals. Note that nitric acid will pit and corrode carbon steel, and hydrochloric acid will corrode stainless steel. For devices possibly contaminated with organics, hexane or acetone may be used; however, thorough decontamination using the water-based method is preferred to minimize both potential cross-contamination in the field and waste-solvent generation.

SAMPLE PRESERVATION AND HANDLING

A sample may require preservation steps to prevent chemical and biological changes that will affect the outcome of the laboratory analysis. To accomplish this preservation or "stabilization," most samples are chilled and pH controlled. However, it is important to note that holding times (i.e., limits on the length of time that a sample may be considered "stable" prior to being analyzed by the laboratory) still apply.

Chilling to 4°C may reduce evaporation of volatiles, absorption, biological action, and hydrolysis of organic chemical compounds, plus reduce or prevent chemical reactions. Chilling is done by placing samples in filled ice chests immediately after the samples are taken. Samples should not be frozen. In the event that "blue ice" or other commercial ice packs are used, take care to ensure that the temperature does not fall below 4°C. A common mistake is using dry ice, which is frozen carbon dioxide at a temperature of −56°C.

Preservation using small amounts of nitric acid to acidify the sample is done to prevent metals from precipitating out of aqueous solutions. The sample is acidified by adding a drop or two of reagent grade 70% nitric acid and reducing the pH to less than 2. Often this is done by the laboratory before sample containers are provided to the customer (prepreserved containers). For certain other types of analyses, other acids or solutions may be used, such as hydrochloric acid or sodium thiosulfate.

QUESTIONS FOR REVIEW

1. Why are materials such as polyethylene inappropriate when selecting containers for sampling organic solvents?
2. What types of compounds affect stainless steel?
3. How do you use a glass thief to collect a sample?
4. What type of sample is a COLIWASA used for obtaining? What does COLIWASA stand for?
5. A soil auger is used to do what?
6. What type of sample is a tube sampler designed to collect?
7. How is a split spoon sampler described?
8. A Geoprobe™ is a device that is designed to collect soil core samples. What are the advantages of this type of sampling equipment?
9. Dedicated groundwater sampling pumps have several advantages. What are they? Discuss them.
10. Why are organic solvents that could be used for decontaminating equipment not used in the field sometimes?
11. What are some potential problems when using materials other than regular ice to cool samples?

ACTIVITIES

1. Obtain a sampling devices catalog or manufacturers information on several types of sampling equipment (references, Lab Safety Supply, Janesville, Wisconsin, or Arts Manufacturing, American Falls ID, or Forestry Suppliers, Jackson, Mississippi). Prepare a brief summary of the capabilities and cost of one soil sampling device and one water sampling device.
2. In the classroom, handle equipment and become familiar with it. Obtain a COLIWASA sampler and a thief. Describe the difference between the two devices.

READING

Test Methods for Evaluating Solid Waste (SW-846), 1993. Environmental Protection Agency National Technical Information Service.

3

Sampling Techniques and Procedures for Soil and Water

Patrick K. Holley

Upon completion of this section, you will be able to do the following:

- Describe grab and composite sampling techniques.
- Recognize and describe various automated sampling techniques and equipment.
- Understand and apply the sampling procedures that are utilized for surface, wipe, sweep, chip, soil, wastes, and groundwater.
- Demonstrate knowledge of the principles of groundwater sampling, including well purging and calculation of well casing volumes.

INTRODUCTION

People who are responsible for sampling often are unaware of the proper techniques for sampling various media. These media may include soil, surface water, groundwater, and sludge, in containers or tanks. Although guidelines do exist for performing this activity, they are generally focused on laboratory analytical procedures. EPA references such as *Test Methods for Evaluating Solid Wastes—SW-846* (EPA, 1993) and *Standard Methods for the Examination of Water and Wastewater* (EPA, 1992), and field experience form the foundation for sample collection, preservation, and laboratory analysis methods. However, it must be recognized that sample collection guidelines are sometimes adapted to accommodate special or unique analysis parameters or sample types.

GRAB SAMPLES

A *grab sample* simply represents a snapshot in time and location. These types of samples are collected from a specific location at a specific interval in time and are used to identify conditions at the time of sample collection. Grab samples are generally used where materials are homogeneous and accessible to a sample device such as a dipper, thief, or bottle.

COMPOSITE SAMPLES

A *composite sample* is a nondiscrete sample composed of several grab samples taken from specific locations and then analyzed to produce an average value. Composite sampling economizes on the number of samples processed by the laboratory and, thus, on the funds required for the project. Individual grab samples are combined in equal proportions in a single container, agitated, placed into an appropriate sample container, and then analyzed. For example, if you needed to assess a large amount of a stockpiled material that is potentially contaminated with mercury, you could composite a series of samples taken from several stockpiles. Figure 3–1 illustrates the principle of compositing sample groups from different batches, strata, or layers. Each individual grab sample is combined to form a composite sample for Stockpile A in Figure 3–1.

Assuming that a sensitive laboratory method is selected for analysis, this will allow you to determine whether mercury is contaminating the stockpile while sending only one sample per stockpile to the lab. It can also indicate that no contamination exists in the other stockpiles with a high degree of certainty. Note that if an insufficiently sensitive laboratory method is selected for the analysis process, the presence of contaminants may be masked because of dilution that results from the compositing process.

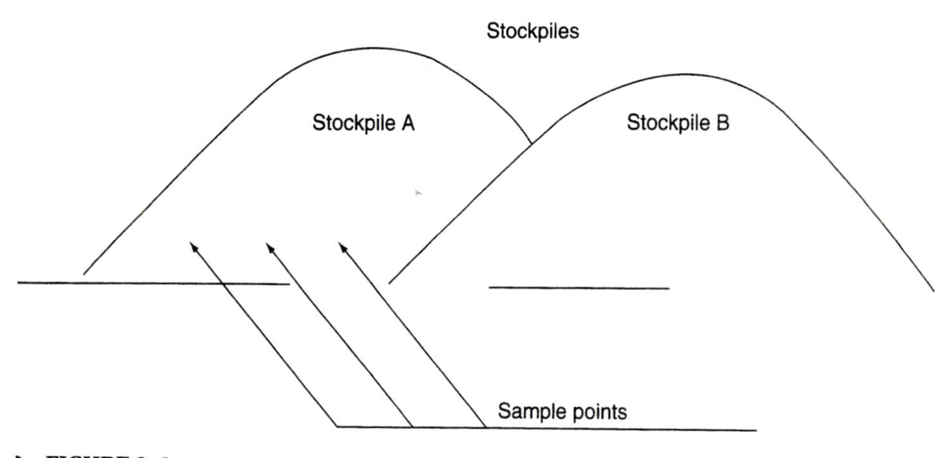

FIGURE 3–1
Composite sampling.

As an example, if the laboratory method's detection limit is 1 mg/kg and the regulatory limit is 5 mg/kg, as a general rule no more than five samples would be composited together to avoid diluting one or more of the five samples below the detection limit. This precludes the possibility of diluting one sample that has a concentration at or near the regulatory limit from escaping detection. If a concentration of 1.2 mg/kg is reported by the laboratory, each of the five individual samples making up the composite could be analyzed to determine which is the source of contamination.

Composite samples also may be collected from the same point at different times. As an example, a continuous wastewater discharge may be sampled using an autosampler that collects identical volumes at scheduled intervals. This type of sampling ensures that average values for the entire sampling interval will be obtained. This method is often used in surveillance sampling procedures employed by regulatory agencies.

In some sampling situations, especially water discharges or flows within a process when flow rates vary significantly over time, a "flow-weighted" sample will be collected. For example, a programmable autosampler would be set to collect 100 milliliters of sample for each 10,000 gallons of flow. The flow may drop to very low levels at certain times of the day and increase dramatically at others, thus requiring a flowmeter input signal to the autosampler to collect samples based on flow intervals rather than timed intervals.

HAZARDOUS WASTES SAMPLING FROM CONTAINERS

To sample drums using a COLIWASA, the valve should be opened fully, and then the tube lowered slowly into the container (count to 5 while lowering the tube) without touching the sides or disturbing the material. Once fully lowered, close the valve mechanism. Withdraw the tube and place the tip into the sample container, and then release the valve, slowly allowing the liquid to flow into the container without splashing or spilling.

Grab samples do not collect different layers or strata; therefore, sampling of tanks and containers is normally performed with a cross-sectional sampling device such as a COLIWASA or glass thief. In this method, the sampling tube is lowered slowly into the tank or drum so that the layers are taken into the tube with the uniformity that exists in the material being sampled. Representative samples from tanks or containers that cannot be cross-sectionally sampled may be taken at the top, bottom, and middle with a device such as a bacon bomb sampler, which can be lowered to specific depths and then a valve can be opened, allowing liquid to enter.

Grab samples may be taken from containers, sumps, tanks or other containers using bailers, dippers, or remotely sealable samplers. Care must be exercised to ensure that there are no trace quantities of hydrocarbons or other chemicals of different solubility or specific gravity, which are predominantly found at the surface or near the bottom of the body.

Sludge on the bottoms of tanks or containers may be sampled with a clamshell-type device (ponar dredge) or a simple penetrating coring device. The coring devices have the ability to obtain a certain depth of the sludge below the uppermost layers.

The clamshell device samples the surface of the sludge or sediment. If the sludge is flowable or very soft, a concave eggshell insert in the end of the device will retain the captured sludge.

SAMPLING SOLIDS AND SURFACES

There are many types of solid matrices we may wish to sample, including soils, powders, granules, smooth and porous surfaces, aggregates, and others. Each material may require varying types of devices to obtain representative samples. Many materials may contain pockets or mixtures of the various solids and may therefore be heterogeneous.

When sampling solids, ask the following questions:

- How deep does the sample need to be?
- What types of solids make up the material?
- How hard or soft is the material?
- Is power-driven equipment necessary?
- Can power equipment reach the site?
- Is volatile organic analysis required?

Soils and Solids

Completely solid materials require chisels or power equipment to remove adequate quantities for analysis. If contamination has penetrated deeply into concrete, for example, a power coring device or hole saw may be used to obtain a representative sample (note that organic contaminants may be volatilized due to friction and heat). To collect a proper sample, you should ask whether the surface or the entire depth of the concrete is the mass of interest. The answer will determine whether a sample is chipped out of the concrete's surface, a core is cut out, or a wipe sample is taken of the area.

Soil augers and tube samplers used for taking soil core samples at the surface or at specified depths also are available as combination barrel augers. These devices are capable of digging down to the sample depth and then collecting a subsurface soil sample by attaching a penetrating core-cutting attachment onto the end of the sampler rod. Before the soil-sampling process is started, a liner is inserted into the penetrating cutter so that the sample, when collected, does not contact the cutter's inner walls. A mallet or slide hammer may be used to drive the cutter into the ground to obtain the core. The inner liner is then removed and caps are placed over both ends of the liner, thus forming the sample container. For VOCs or other organic analysis, a 3- or 4-inch-wide piece of Teflon tape is placed over each end of the sample tube before placing plastic caps over the ends to seal the sample more securely. These liners or tubes are normally constructed of brass or stainless steel.

The core sample may be taken with the tube sampler in the following manner:

1. Attach the auger bit to a drill rod extension, and then attach the *T* handle. Use Teflon tape on the threads to prevent seizing.
2. Clear the area to be sampled of any surface debris (e.g., rocks, twigs, or litter). When taking subsurface samples, you should remove the first 3 to 6 inches of soil in an area approximately 6 inches in radius around the drilling location.

3. Begin augering, periodically removing and depositing accumulated soils onto a tarp or plastic sheet spread near the hole. This prevents accidental brushing of loose material back down the borehole when removing the auger or adding drill rods. It also facilitates refilling the hole and avoids possible contamination of the surrounding area.
4. After reaching the desired depth, slowly and carefully remove the auger from the boring. When sampling directly from the auger, collect a sample after the auger is removed from the boring and proceed to step 10.
5. Remove the auger tip from the drill rods and replace with a pre-cleaned tube sampler in which a sample sleeve has been inserted.
6. Carefully lower the tube sampler down the borehole. A slide hammer is placed over the drill extension rod, and the tube sampler is hammered into the soil to collect the sample.
7. Remove the tube sampler and unscrew the drill rods.
8. Remove the threaded top of the tube sampler and remove the sample collection sleeve from the tube.
9. Place plastic caps on each end of the sample sleeve and mark the sample appropriately.
10. If volatile organic analysis is to be performed, place a piece of wide Teflon tape over each end of the sample sleeve prior to putting on the plastic caps.
11. If another sample at a greater depth is to be collected from the same hole, reattach the auger bit to the drill rod and follow steps 3 through 10, making sure to decontaminate the auger and tube sampler between samples.

Soil Sampling for Underground Tank Removals

Upon removal or closure of underground tanks in which motor fuel has been stored, regulations often stipulate that you must determine the level of contamination in the underlying soils. Samples must be collected carefully to avoid erroneous laboratory results. Incorrect characterization of contamination levels may lead to excess removal of soils or inadequate removal. This could affect both the cost and effectiveness of the remediation process. Such samples must be collected in the following way:

1. Determine the sample location or locations. Small tanks of less than 1,000 gallons may require only one sample; 1,000 to 10,000 gallons may require two samples; more than 10,000 gallons may require three or more. Other details regarding sample locations may be found in the LUFT manual (1989).
2. Ensure that safety procedures are followed when sampling in an excavated area (i.e., ladders, CGI and O_2 testing, and PPE).
3. Locate the sample area and remove loose or dried surface soil with a clean spatula or trowel until undisturbed soil is reached.
4. Put on a clean pair of gloves.
5. Using a stainless steel sampling sleeve, push the sleeve into the soil as far as possible or until the sleeve is full and then move the sleeve gently side to side and remove the sleeve containing the soil sample.
6. Wrap the ends of the sleeve with wide Teflon tape and secure with plastic end caps.

7. If the soil is difficult to penetrate manually, a tube-type sampler containing the stainless steel sleeve may be used.
8. It is important to note carefully the exact location of samples, including depth, on sample collection forms.
9. The sample should be marked and labeled as appropriate, immediately placed in a cooler at 4°C, and transported to a laboratory as soon as possible.

Cross-contamination often occurs in this type of sampling when the sampling technician collects the actual petroleum product from the tank to confirm the source of the soil contamination. Do not place this product sample in the same cooler as the soil sample. Preferably, the product should not be handled at all by the same technician on the same sampling effort.

Surface Wipe, Chip, or Sweep Samples

In some cases when the contaminant is present only on the surface of the structure, wall, floor, or equipment, wipe samples may be collected. These samples may be obtained from solid or relatively nonporous surfaces by wiping a specified area, normally 100 cm^2, with an absorbent pad saturated with an appropriate solvent. Examples include hexane solvent for PCBs and deionized water for heavy metals.

The wipe sample may be taken using the following general procedure:

1. Choose the appropriate sample points. Prepare and label wipe sample kits, containers, and templates. Measure and document the sample location on a map or a photo.
2. To facilitate later calculations, record the surface area to be wiped.
3. Don a new pair of chemical-resistant gloves.
4. Remove the soaked swab from the wipe sample container.
5. Place the template on the surface sample location.
6. Wipe the surface outlined by the template first vertically in an *S* pattern without lifting the swab and then horizontally in the same *S* pattern.
7. Place the swab back into the wipe sample container and seal it tightly. Use a Teflon seal for organic wipe solvents.
8. Mark the area sampled in some way or leave the template taped to the surface so that the site may be reviewed and sample values compared to actual sample locations.

Chip samples may be taken in cases when a porous surface has been contaminated. These samples are gathered using a chisel and hammer, or an electric impact chisel or hammer. The following general procedure may be used:

1. Choose appropriate sampling points. Prepare and assemble chisels, sample containers, and other supplies. Measure off the designated sample point and document it on a drawing or a photo.
2. To facilitate later calculations, record the surface area to be chipped.
3. Don a new pair of chemical-resistant gloves.
4. Open the container holding the laboratory-cleaned chisel or equivalent sampling device.

5. Chip the area designated, first horizontally, and then vertically, to an even depth of approximately ⅛ inch.
6. Place the sample chips in an appropriately prepared sample container with a Teflon-lined cap for organic analytes.

Sweep samples may be collected for dispersed solids and bulk granular or powdered materials. These samples are collected using a contaminant-free brush or broom and pan to sweep and collect material from the specified areas. (Composite sampling is often employed with these dispersed bulk materials.) Samples then are placed in appropriate containers, labeled, and transported to the laboratory for analysis.

Sampling Surfaces for Lead Dust

Lead residue may accumulate in dust on surfaces where lead-based paint is present, due to wear and abrasion of the paint. To determine the lead content, select an area to be sampled, and then clean your hands with wet wipes carried for that purpose. A template also should be cleaned upon collecting each sample or, alternatively, new templates used each time. Use the wet wipes as swabs for collecting the wipe sample. (The wipes should contain as little fragrance, oils, etc., as possible.)

1. Put on a clean pair of rubber gloves.
2. Pull a wipe from the wipe container, fold several times, and place into a sample tube (plastic centrifuge tube). This is the field blank.
3. Pull a wipe from the container, fold appropriately, and wipe a sample area of 1 square foot.
4. The wiping technique consists of wiping the sample area in an *S*-shaped pattern, trying to achieve as close to 100% coverage as possible. An additional wiping of the area should be performed at 90 degrees to the initial wipe.
5. Each wipe should be placed in an appropriately labeled tube immediately after the wiping process. The wipe should not be allowed to touch any surface, including the template (if one is used), other than the surface to be wipe-sampled.

WATER SAMPLING

Water samples are collected from many diverse sources, such as rivers, ponds, tanks, pipelines, reservoirs, sumps, groundwater wells, and treatment plants. Collecting representative samples in each of these environments is critical. The containers and sample devices vary greatly—and have been discussed previously—but should be selected to match the requirements of the analytical method to be employed and the constituents of interest.

Surface Water

Surface waters or open water bodies may be sampled directly using grab samples. If sampling rivers or streams, make sure to always take samples from upstream. Avoid disturbing any sediment or substrate. Submerge a bottle underneath the water's surface, allow it to fill, and then cap it. Remember that if you are in a boat with a gasoline engine, take care to sample from the bow aiming upstream because fuel and engine oil are sources of cross-contamination. Also be careful not to place collected samples near fuel or oil.

The dipper also may be used to collect water surface samples from tanks, processes, treatment plants, and so on. Make sure to collect the sample from the discrete location that has been selected. For example, if an outfall discharge pipe that pours into a settling tank is the desired sampling point, the dipper should be placed directly into the flow of the discharge. This avoids the influence of the static water body, which could make the outfall sample unrepresentative.

Drinking Water Sampling for *Coliform* Bacteria

When sampling water supplies, it is necessary to collect a sample after running the faucet or outlet to obtain a representative sample. Sample containers have been prepared in the laboratory, contain a preservative, and are sterile, so do not use unsterilized containers and do not rinse the containers.

The following steps should be used when sampling for *Coliform* bacteria.

1. Pick a water tap that is commonly used. Do not use swivel faucets or outlets that have filters, aerators, or other such attachments. Remove these attachments if necessary.
2. Clean the faucet with rubbing alcohol inside and out using a sterile swab or gauze pad.
3. Open the water faucet or valve to a full, steady stream.
4. Let the water run for 10 minutes.
5. Hold the sample container near the bottom so that when inserted into the water stream it is not forced out of your hand.
6. Reduce the water flow.
7. Open the sample container, carefully insert the opening into the water stream, and fill the container. Leave some air space at the top.
8. Recap the container and transport it to the lab within six hours. This sample has a relatively short holding time within which the analysis must be started.

Sampling Procedure for Inorganic Chemicals in Water

Water may be sampled to determine the concentrations of inorganic chemical contaminants such as zinc, arsenic, nickel, copper, and many others. This procedure also requires that the water source be run for 10 minutes to obtain a representative sample. Do not rinse out containers. Containers have a small quantity of acid added to preserve the sample.

To sample for inorganic chemicals in water, follow this procedure:

1. Pick a water tap that is commonly used. Do not use swivel faucets or outlets outfitted with filters, aerators, or other attachments. Remove these attachments if necessary.
2. Open the water faucet or valve to a full, steady stream.
3. Let the water run for 10 minutes.
4. Hold the sample container near the bottom so that when inserted into the water stream, it is not forced out of your hand.

5. Reduce the water flow.
6. Open the sample container, carefully insert the opening into the water stream, and fill the container. Leave some air space at the top.
7. Reseal the container and transport it to the lab for analysis. Holding time is up to six months on these samples.

A variation on this procedure when sampling for lead in drinking water is to allow the sample to be collected from the faucet or valve without first running the water. This captures maximum concentrations that may have built up in the piping over time. Water should stand overnight before being sampled.

Sampling Water for Volatile Organics Analysis

When collecting samples for volatile organics analysis, several items should be considered. Whether the sample is collected from an open water body or from a valve or tap will determine how it is handled. For example, if collecting a sample from an open water body, a wide-mouth 1-L container or beaker is used to collect the sample for transfer into the VOA vial.

The procedure for sampling water for volatile organics analysis follows:

1. Open the water tap or valve and allow to run for 10 minutes to clear water lines and ensure a representative sample.
2. If sampling an open water body, dip a 1-L container into the water and collect the water for transfer into a VOA vial.
3. Fill the VOA vial to just overflowing (where a meniscus has formed) without creating air bubbles or excess agitation.
4. Place the container on a level surface.
5. Place the septum, Teflon side down, on the top of the convex sample meniscus.
6. Seal the sample with the screw cap.
7. Invert the container and lightly tap the cap on a solid surface.
8. Be sure there is no trapped air in the container.
9. If air bubbles are observed, open the cap, refill just to overflowing, cap, and check again for air bubbles.

Groundwater and Subsurface Sampling

Groundwater and subsurface sampling is often a component of site characterization when contaminants have migrated below the soil's surface. Determinations must be made on the location and type of subsurface soil samples (usually accomplished with soil borings) and subsequent placement of groundwater wells. Hydrogeological conditions at the site must be considered when preparing such a sampling program. Characterization of the groundwater flow and any possible chemical interactions that may occur between the analytes and the subsurface matrix must be completed to place wells and determine their depths.

Wells may be sampled using many devices; however, certain procedures for removing water that has accumulated in the casing must be followed. This allows fresh groundwater to flow into the casing so that a representative sample may be taken. Most common groundwater well casings vary from 1½ to 6 inches (some may be larger) in diameter, with a slotted or screened section of pipe at the particular depth interval from which the samples are being taken. Wells are sealed with bentonite clay (usually above the screen) to prevent the downward migration of water and contaminants, and are filled with coarse sand around the screen's exterior to filter out groundwater particulates.

Wells must be purged prior to sampling. Wells with slow rates of recovery (i.e., flow through the casing) generally require the removal of three to five well casing volumes prior to sampling. Wells with fast recovery rates may be sampled after ten well volumes have been removed. The procedure is as follows:

Procedure

1. Log in your start time on the field data sheet.
2. Remove the well cap and measure the depth of groundwater and the total depth of the well.
   a. Measure for the predetermined mark located on the PVC casing. If there is no predetermined mark, make one on the inside or outside of the PVC casing for consistent measurements in the future. Measuring from one defined spot ensures consistency, since the PVC is usually cut unevenly.
   b. Turn on the depth-to-groundwater meter and lower it until it beeps. Record the depth to the groundwater.
   c. Measure the total depth of the well by sinking the meter down to the bottom of the well. Be sure to turn off the meter so the monitor won't continue beeping. Once the meter has hit the bottom, measure the depth, and add 0.33 feet to the total depth. (The additional 0.33 feet is the length of the weighted tip below the conductivity sensor.)
   d. The height of water in the well is calculated as follows:

      Water column (ft.) = (Total well depth) – (Depth to groundwater)

3. Calculate the well volume as follows:

   Casing volume (gal.) – (Water column) × (Casing variable)

   The *casing variable* is used to calculate the number of gallons of water held in a particular size of casing: for example, 2-inch schedule 40 pipe, which is 0.667 gallons/ft. (The table of casing values is located in the bottom left corner of the groundwater well-sampling sheet in Figure 2–2.)
4. For testing purposes, EPA guidelines state that a monitoring well should be purged (well water removed) a minimum of three to five times the volume of water in the well casing. The purpose of this purging is to remove stagnant water. This will be demonstrated by consistent pH, electrical conductivity (EC), and temperature readings.
   a. To calculate the minimum amount of water that should be purged, multiply the casing volume by 3.
   b. The total volume of water, in gallons, that will be sampled is then divided up into three or more field measurement points. For example, if the casing volume is 1.6 gallons, multiply by 3 to get 4.7 gallons. Normally, with 4.7 gallons, field measurements would be taken at the 2-, 3-, 4-, and 5-gallon marks.
   c. Calibrate the pH and EC meters (see instructions with the analytical meters for specific calibration procedures). Document calibrations in the logbook.

   d. After the meters are calibrated and the sample is ready to be tested, write down the time the sample was obtained and then measure the water temperature. The order of the field tests is as follows: temperature, pH, and electrical conductivity. (If using a Hydac meter, make sure that it is set at the correct temperature since the EC measurement depends on temperature).
   e. The routine field measurements are temperature, pH, and EC. On occasion, turbidity is required.
5. After making the field measurements, collect the appropriate sample(s) for the laboratory. Laboratory samples are taken after the well has been purged of at least three times its volume of water, or until stable pH, temperature, and conductivity are observed. Stability is defined as a difference of less than a 10% for temperature and EC over successive well volumes. If stability is not reached within ten well volumes, it may be necessary to sample at that time. This is to ensure a representative sample of groundwater. The requirements of the client will dictate the kinds of sample containers to use. Clearly label all samples and ensure that they are stored properly.
6. If there are multiple monitoring wells, a new disposable bailer must be used for each well. If a 6-foot reusable bailer is used, it must be decontaminated with Alconox soap and a tap water rinse, followed by deionized—purified by removing ions (DI)—water rinses before using on a different well. Collect an equipment blank between each use of the decontaminated reusable bailer. Reusable bailers should be used only for purging. Use new disposable bailers for actual sampling.

Chemical parameters may be measured in the field to determine whether adequate purging has occurred. These include pH, specific conductance, and temperature. Field instruments may be used to sample the purge water, and when these parameters stabilize (i.e., vary less than 10%), this should indicate that the water in the well is representative of the water contained in the underground formation or aquifer. See Figure 3–2 for information on collecting groundwater samples. Well-casing volume calculation factors also are listed for various casing sizes.

QUESTIONS FOR REVIEW

1. What is meant by grab sampling?
2. Composite samples may be collected based on individual grab samples and then combined in equal proportions to form the sample. What are some advantages of composite sampling?
3. What is an autosampler used for? Under what type of sample collection conditions may it be used?
4. Layers or strata may exist in certain types of samples. What are some of the methods for collecting representative samples of these materials?
5. What must be considered when establishing sampling plans for solid materials?
6. What is the basic procedure for obtaining a soil core sample?
7. Groundwater may be sampled once certain parameters have stabilized or when a certain volume has been removed from the well. What are these parameters and volumes?
8. What is the well-casing volume for a well 32 feet deep and 5 feet to groundwater, with a 3-inch diameter schedule 40 pipe?
9. What are some items that should be documented on well-sampling logs or forms?
10. What are some potential cross-contamination problems that may be encountered when sampling groundwater?

GROUNDWATER WELL SAMPLING SHEET

Project Name: ______________________

Project #______________ Sampler(s): ______________

Date:______________ Sample Point:______________

Time of Start:______________ Water Level Measurement Method: ______________

Depth to Groundwater:______________ Purging Method: ______________

Measuring Pt. Description:______________ Construction Material: ______________

Well Depth:______________ Sample Method & Construction: ______________

Material: ______________

Was Well Depth determined by measurement or reference? ______________

Well Volume:

Total depth – depth to groundwater = water column (ft.)

Water column × multiplier for casing = casing volume (Diameter see chart, lower left)

Well Number											
Time											
Volume purged (gallons)											
Purge Rate (gpm)											
pH											
Temperature (Celsius)											
Conductivity (micromhos/cm)											
Turbidity (NTU)											
Color											
Odor											
Other											

Volumes per unit length for common well-casing diameters

Pipe size	Gal/Ft		L/Ft	
Inches	Sch* 40	Sch 80	Sch 40	Sch 80
1.5	.106	.092	.401	.348
2.0	.174	.153	.658	.579
3.0	.384	.343	1.45	1.30
4.0	.661	.597	2.50	2.26
6.0	1.50	1.35	5.68	5.11

* Sch = schedule, a measure of pipe wall thickness.

FIGURE 3–2

ACTIVITIES

1. Study procedures required for collecting various types of samples and assemble necessary supplies and equipment. Find an acceptable location and set up areas to collect mock samples. Collect one soil sample and one water sample.
2. Take a field trip to a local chemical plant or waste water treatment plant and its laboratories. Prepare a brief report summarizing sampling and analysis activities at the lab.

READINGS

Standard Methods for the Examination of Water and Wastewater. 17th ed. Environmental Protection Agency.

Test Methods for Evaluating Solid Waste (SW-846), 1993. Environmental Protection Agency National Technical Information Service.

4

Sampling Plans

Patrick K. Holley

Upon completion of this section, you will be able to do the following:

- Identify and describe the key elements of a sampling plan, including legal requirements, sampling objectives, known contaminants and locations, sample collection, number and type of samples, required analysis, health and safety procedures, and quality assurance and control.
- Develop and prepare a written sampling plan.
- Understand and apply quality assurance procedures.

INTRODUCTION

One of the most important aspects of successful sampling projects involves the preparation and development of a sampling plan. The development of a sampling plan that is legally and scientifically defensible is an overall objective. This will be accomplished by preparing a written plan that outlines the sampling objectives, number and type of samples, sampling equipment, and safety procedures, among other elements. Once this planning activity is performed and reviewed by appropriate individuals, a great many potential problems and obstacles will have been anticipated and addressed. This plan may be simple or complex, depending upon the sampling site, but it is important that it be written so that procedures, methods, and assumptions may be validated as necessary in the future.

PLANNING

Planning is required to ensure that safety rules are adhered to and that representative, uncontaminated samples are taken. Prior to taking samples, all necessary paperwork, containers, protective clothing, sampling devices, and transportation should be obtained.

Other considerations include having approval to enter the site, providing decontamination facilities for people and equipment, and making arrangements with the analytical laboratory.

SAMPLING PLAN COMPONENTS

A sampling plan is developed during the planning process to address many considerations described in this section. These include stating the objectives of the sampling effort, determining the possible chemical analysis parameters for each sample, following the appropriate health and safety procedures, using the analytical and sampling methods, determining the methods to obtain the appropriate number of samples to provide adequate results, and establishing a minimum number of samples to meet statistical criteria.

The sampling plan should be developed with the assistance of laboratory professionals who understand that establishing clear criteria is crucial in developing sampling plans that will stand up to regulatory scrutiny. The sampling plan may be modified as analytical laboratory results are obtained. It may become more detailed or complex due to new findings that redirect the sampling effort. Be ready to modify the plan as necessary to accommodate regulatory agencies, legal requirements, and changing environmental conditions.

Address the following items in the plan:

- Objectives of sampling.
- Legal requirements.
- Known contaminants and/or location of each contaminated area.
- Sample collection procedures.
- Number and location of samples.
- Required analyses for each sample.
- QA/QC procedures.
- Health and safety procedures (plan).

These items will be expanded on and discussed in this chapter.

Sampling Objectives

One of the most important sections of the sampling plan is the section on objectives. The objectives must be developed and written clearly so that other components of the plan may be properly defined. As an example, one objective of a sampling plan may be to sample and test drums to determine whether hazardous waste characteristics or criteria are met. This clearly requires that properties and analytes of concern be selected from Resource Conservation and Recovery Act (RCRA) regulations found in the Code of Federal Regulations (CFR) 40, Part 260.11. If wastewater discharge from a permitted treatment plant is to be tested for adherence to NPDES permit conditions, then such things as constituents of concern and methods are identified in the permit and NPDES regulations. These sampling goals should be listed within the sampling plan.

Legal Requirements

The definition of legal requirements complements the previous discussion on sampling objectives. A legal requirement is the clearly identified law, regulation, or standard that is the driving reason or justification for the sampling program. Examples of legal requirements include 40 CFR Part 761 for PCB cleanup standards, SW-846 (EPA, 1993) for hazardous waste determinations and classification, and *Standard Methods for the Analysis of Water and Wastewater* (EPA, 1992) for sampling and classification of water discharges from wastewater treatment plants. This legal basis should be determined and identified within the sampling plan.

It also is legally required that proper approval for access to the site be obtained. This should be done far in advance of the date that the sampling activity is to begin so that no delays are imposed on the project. Most often, obtaining site approval is simply a matter of getting a letter of permission from the property owner/operator. In cases of hazardous waste sites, a prime contractor is usually managing all programs for an owner or group of owners, so access to the site requires coordination with this contractor. In the case of abandoned or condemned property, city or county ordinances and state property laws should be consulted to determine who has the authority to grant access. All of these authorizations should be obtained and included in the sampling plan, and provided to sampling crews for their use on-site.

Known Contaminants and Locations of Contaminated Areas

The knowledge of people employed at the contaminated facility will often form the basis of the sampling plan. Knowledge such as where hazardous materials have been stored or spilled in the past and chemical properties and characteristics will enable those preparing the plan to identify sampling parameters. Interviews with employees (past or present) may provide more detail. Research on products or substances handled, including obtaining material safety data sheets or product technical information sheets, also will provide details on chemicals potentially present. Prior or historical sampling results often are available and may indicate which areas are contaminated and what contaminants are present. A review of past uses of areas of interest conducted by studying aerial photos, archived engineering drawings, and plot plans also may yield substantial information on where and how contaminants may have been introduced into an area. City or county historical records such as photos, maps, planning drawings, land use permits, and hazardous materials permits also may provide valuable information.

Sample Collection Procedures

Exact collection procedures should be written in a step-by-step fashion in the sampling plan. Include all necessary details (specific container specifications, preservation method, sampling devices, etc.). This allows technicians who will collect samples to review, study, and reference all procedures. Do not rely on memory for proper collection of samples.

Number and Location of Samples

As discussed earlier, sample selection methods include several ways to determine the number and location of samples, among them judgmental (also known as *authoritative*), simple random, stratified random, and systematic random methods. The *random method* requires the area to be divided into numbered areas and then either a random number generator or a random number table is used to select the individual numbered areas to sample. The limitations of this method are that it is subject to many variables, so its validity may be questioned based on any of those.

Statistical methods, however, also must be employed carefully, since many environmental sample analytes are not normally distributed. A limitation of methods utilizing statistical bases is that preliminary samples must be collected and results of laboratory analyses obtained to be able to employ statistical methods on a data set. The basis for sample selection should be prepared in the written sampling plan and reviewed by laboratory or environmental professionals who understand these limitations.

Required Analyses for Each Sample

Selection of laboratory analysis methods and analytes of concern should be made in the context of the sampling objectives and legal requirements discussed earlier. Consult the Sample Collection, Preservation, and Laboratory Method Chart on pages 4–6 in the section on sample preservation and handling in addition to source procedure manuals such as *SW-846* (EPA, 1993) and *Standard Methods* (EPA, 1992). When reviewing a sampling plan for inclusion of analytes, the following factors should be considered:

- Historical or preliminary survey data indicating constituents of concern.
- Regulatory agency input—agencies often request additional analytes of interest due to local conditions or adjacent sites that are adding or contributing contaminants to a site.
- Regulatory limits or thresholds—selection of laboratory methods depends on the levels of detection required. Lower detection limits generally require more sophisticated instrumentation and accompanying methods.
- Sample matrices—there are different methods based on the makeup or matrix of the sample.

Quality Assurance/Quality Control Procedures

The validity of field sampling techniques is ensured by a quality assurance program and quality control procedures. Quality assurance (QA) is the total program designed to ensure the reliability of collected samples and their analytical results. Quality control (QC) is the routine application of procedures and methods such as duplicate samples, split samples, blank samples, and spike samples that may be used as appropriate. These methods may be applied to field sampling operations and the selected analytical laboratory. Both the laboratory and the sampling operations require QA programs. The following procedures help ensure the validity of field sampling techniques:

- A field blank sample may consist of pure water or other pure substance prepared in the field that is run through the same handling and laboratory process as the actual samples. The purpose of this type of sample is to verify that cross-contamination is not occurring in the field and that lab results are consistent with the absence of the contaminant. At least one blank should be run with each sampling event or once each day, whichever is more frequent.
- A travel or trip blank is a sample that the laboratory has prepared using purified water or other appropriate substances, taken to the field with the other sample containers but not opened or otherwise handled. The purpose of this type of sample is to verify that field procedures are not contaminating containers or samples. The laboratory verifies that internal procedures produce values consistent with a blank sample.
- Sampling technicians prepare a rinseate or equipment blank by running deionized or purified water through a device or piece of equipment used in the field for sampling *after decontamination but prior to sample collection.* This rinse water is then sent to the lab for analysis to determine the effectiveness of decontamination procedures.
- A field duplicate is simply a repeat of the sample that is sent to different laboratories for cross-verification or to the same laboratory to see whether the original sample results can be repeated.
- A spike is a sample that is "spiked"—that is, the constituent of concern is added to the sample container to verify recovery rates of the analyte. In some cases this also is done to provide independent verification of the analyte concentrations.
- A split sample is simply a split quantity of the same sample that is provided to another party, such as a regulatory agency, that will send it to its own laboratory. This pair of split samples also may be submitted to the same laboratory to determine the repeatability of the results.

These special samples are sometimes not identified as blanks or QA samples. Instead, the laboratory is provided these samples "blind" so that they receive the same treatment as all other samples. One exception to this practice is that if a high concentration of analyte (i.e., a "hot sample") may be present, the laboratory should be warned so that analytical instrument detectors are not overloaded.

A sample list should be prepared that identifies the total number of QC samples, spikes, blanks, and duplicates, along with actual samples. The QA program and QC procedures should be documented within the sampling plan.

HEALTH AND SAFETY PLAN

A site-specific safety and health plan must be prepared and kept on site when engaged in operations on contaminated portions of sites listed by federal, state, or local agencies as requiring cleanup or mitigation. Regulations addressing these topics are found in 29 CFR 1910.120. Safety and health plans are necessary to establish consistent procedures for employees, contractors, and other outside personnel who may be required to enter contaminated portions of the site to perform various duties such as sampling, drilling, excavation, demolition, air monitoring, and agency inspection. The elements of a site safety and health plan address site control, decontamination, air monitoring, and other items discussed in Chapter 6.

QUESTIONS FOR REVIEW

1. What are some of the goals of a sampling plan?
2. Legal requirements are included in sampling plans to address such items as regulatory limits and protocol established in government regulations. Why is it important that these be identified in the sampling plan?
3. What are some methods that may be used to determine how to obtain legal access to a site that is to be sampled?
4. What are some ways to gather information on the materials or constituents of interest at a particular site?
5. Why are all sample collection and handling procedures included in the sampling plan?
6. What objectives does a quality assurance program seek to accomplish?
7. Quality control procedures exist for the purpose of implementing the quality assurance program. What are some typical examples of such procedures?
8. Why is it important to state the objectives of the sampling plan?
9. In what ways may maps, drawings, plot plans, and other visual representations be useful?
10. What are the sources of information regarding site use, hazardous materials, and previous uses on the local government level?

ACTIVITIES

1. Given a facility that has three separate waste water discharges (one contaminated with lead, one contaminated with 1, 1, 1–trichloroethane, and one contaminated with oil and grease), develop a sampling plan. Each discharge pipe has a sample point with a hand-operated valve.
2. As a second part of this sampling plan, develop a cost estimate for the sampling technician time and transportation as well as laboratory charges for each sample. This may be based on a laboratory rate sheet or estimate from other sources.

READINGS

Standard Methods for the Examination of Water and Wastewater. 17th ed. Environmental Protection Agency.

Test Methods for Evaluating Solid Waste (SW-846), 1993. Environmental Protection Agency National Technical Information Service.

TECHNICAL REFERENCES

40 CFR 260.11.
29 CFR 1910.120.

5

Documentation and Data Interpretation

Patrick K. Holley

Upon completion of this section, you will be able to do the following:

- Identify, describe, and complete proper documentation for sample collection efforts, including field notes, chain-of-custody forms, and a sample log or request.
- Understand and apply sample data interpretation methods and concepts.
- Understand and interpret laboratory analytical reports.

SAMPLE DOCUMENTATION

Proper documentation of all sample collection activities contributes to the overall quality of data obtained from analytical laboratories. The cost associated with conducting the sampling and subsequent laboratory analysis represents a significant amount. The money and effort spent to obtain these data can be rendered useless if proper documentation is not prepared and maintained. For each sample taken, only a few minutes are required to complete the appropriate documentation. When documents are prepared properly, this information allows questions to be answered, collection dates and times to be validated, good quality control procedures to be demonstrated—and shows others that you and your team know what you are doing!

Field Notes

The sampling technician should maintain a field log book containing all information pertinent to the field sampling program. The technician should make entries in ink in the bound notebook with consecutively numbered pages. If errors are made, the technician should draw a single line through the entry and initial it. The following information should be included:

1. Date (including year) and time (including total time spent).
2. Purpose.
3. Client, including name, address, and contact person.
4. Type of sample(s) and matrix.
5. Number of samples taken, along with sampling point, time sampled, and what type and volume of preservative (when applicable).
6. Description of sample point.
7. Site sketch and field observations, where applicable.
8. Any field measurements.
9. Field instrument calibrations.
10. Condition of sample (preservation, temperature, air bubbles in VOAs, etc.).

Chain-of-Custody Forms

Completed chain-of-custody forms should accompany the samples. *Chain of custody* is the process of tracking a sample's possession from the time the sample is taken until it is processed by the laboratory. The chain-of-custody record must be signed by each individual involved in the process. This also verifies that the sample has not been tampered with or altered in any way. The samples are to be under lock and key or secure seal under the control of one individual at all times.

These forms should contain the following information:

1. Project site.
2. Sample identification number.
3. Date and time of sample.
4. Location of sample site.
5. Type of sample (soil, water, etc.).
6. Signature of sample collector.
7. Signatures of those who relinquish and those who receive the samples, and date and time that samples change possession.
8. Inclusive dates of possession.

See the sample form in Figure 5–1.

Sample Collection Forms

A sample analysis request form must accompany the samples delivered to the laboratory. The sampling technician should enter the following information on the form:

1. Project site.
2. Name of sample collector.
3. Sample identification numbers.

FIGURE 5–1
Sample chain-of-custody form.

Sample Custody			Lab #:
Client:		Contact name:	Phone #:
Address:			Proj. mgr.:
Billing address:		P.O. #:	Fax phone:
Project:	Project name:	Sampler (print and sign name):	Due date:

Date sampled	Time sampled	Matrix	Container amt./type	Preserv-ative	Sample description/Site	Lab#	Comp. or GRAB	Analysis requested					Remarks

Relinquished by	Date/Time	Received by	Relinquished by	Date/Time	Received by

Container Types:		Matrix:	Remarks:
AL	Amber Liter	AQ	Nondrinking Water
AQL	250 ml. Amber		Digested Metals
AHL	500 ml Amber	WW	Waste Waters
Pt	Pint (Plastic)		Nondigested Metals
HG	Half-Gallon (Plastic)	FE	Low DLS; Final Effluent
SJ	Soil Jar	DW	Drinking Water
B4	4 oz. BACT	Sl	Soil/Sludge/Solid
BT	Brass Tube	AR	Air
VOA	VOA (40 ml)	IW	Irrigation Water
OTC	Other Type Container		Nondigested Metals

4. Location of sample site for each sample.
5. Type of sample (soil, water, etc.).
6. Analysis requested (such as analyte [i.e., total PCBs], method, or desired method detection limit.)
7. QC requirements (duplicates, lab blanks, lab spikes, etc.).
8. Special handling and storage requirements (i.e., special turnaround time).

See the example form in Figure 5–2.

INTERPRETATION OF RESULTS

Many factors affect the development of conclusions drawn from analytical laboratory data. Hypotheses should be proposed as the data from analytical laboratories is reviewed. Patterns may become apparent as data are tabulated and presented in graphical form. For this reason, graphical representation is a very useful tool in analyzing sample results. Data that fall outside statistical norms should be reviewed carefully to determine hypotheses as to why they appear to be irregular.

As an example, assume that perchloroethylene (a solvent commonly used in the dry cleaning and manufacturing industries) was detected in the basement of a building located adjacent to a former dry cleaning business. This information resulted in the hypothesis that the contamination originated from the dry cleaner. However, later it became known that a manufacturer several blocks away used perchloroethylene and periodically discharged contaminants into the storm drain that runs near the basement of the building in question. Such considerations must be reviewed carefully before making interpretations that may be incorrect.

Simple histograms, line, and bar charts provide useful ways of viewing data graphically. More sophisticated graphical means also may be used, particularly for subsurface sampling, such as 3-D in computer software modeling programs. Statistical methods for reviewing the data may include standard deviation, mean, median, and variance.

Common explanations for data that are outside of expected values include sample contamination, poor handling and preservation, different analytical methods (i.e., differing sensitivities), different labs, different detection limits, lack of representative sampling techniques, and analyte degradation. These potential explanations for irregular data should be evaluated carefully when proposing hypotheses.

Many complexities that require conclusions to be based on the assembled expertise of individuals practicing different specialties exist on contaminated sites. Review sessions should be held at regular intervals, particularly on complex or large site investigations, to allow these interdisciplinary teams to discuss the data and conclusions being considered.

Date:____________

Time:____________

Sample Collection Log

Sampled by:________

Project name/Site:____

Sample no. ____

Sample location: ____

Composite: ______ YES ______ NO

	Containers used	Amt. collected
Depth of sample: ______________________	____________	__________
Weather/Conditions: ____________________	____________	__________
Requested turnaround time: ______________	____________	__________

Comments/Sketch:

FIGURE 5–2
Sample collection log.

QUESTIONS FOR REVIEW

1. Field notes may be described as a working record of tasks performed by a person collecting samples. Why are field notes important?
2. What are some of the items that should be included in field notes?
3. How should entries be corrected in the notes?
4. What type of notebook should be used when entering information in a field note format?
5. *Chain of custody* refers to a series of individuals responsible for maintaining and handling a sample in the proper manner. Why is it important that this chain of custody be documented?
6. What are some of the items that should be included on chain-of-custody forms? Discuss them.
7. Why is it important to list the date and time that samples were received and relinquished?
8. Sample collection or analysis request forms identify the samples by number and where they were obtained, as well as many other information items. Why is it important to properly number and identify a series of samples collected during the day?
9. What is the importance of information on the sample type or matrix?
10. Information on the laboratory analysis method is included in sample collection forms. Why is this important? Give an example of a laboratory analysis method and explain.

ACTIVITIES

1. Using copies of the form in Figure 5–1, provide all required information for samples taken in Activity 4–1.
2. Obtain from the local drinking water provider (public works or other agency) a current report on measured levels of contaminants in the drinking water. These analyses are required for all public water systems and are available to the public. Discuss them in class.

READINGS

Standard Methods for the Examination of Water and Wastewater, 1992. 18th ed. Environmental Protection Agency.

Test Methods for Evaluating Solid Waste (SW-846), 1993. Environmental Protection Agency National Technical Information Service.

6

Site Safety and Health Plans

Patrick K. Holley

Upon completion of this section, you will be able to do the following:

- Identify and describe the basic components of a site safety and health plan.
- Describe and apply the principles of site control, decontamination, air monitoring, and action levels.
- Describe and apply appropriate levels of personal protection to be utilized during sampling activities.

SITE SAFETY AND HEALTH PLAN COMPONENTS

A site-specific safety and health plan should be developed for sampling personnel to follow, including guidelines and instructions on protective clothing and equipment, emergency response, confined space entry, and the following procedures:

- A safety and health risk or hazard analysis for each site task and operation found in the work plan.
- Employee training assignments to ensure that the employee understands and has specific knowledge of responsible safety and health personnel, hazards present on the site, use of PPE, work practices to minimize exposure, engineering controls, medical surveillance requirements, and contents of the site safety and health plan.
- Personal protective equipment to be used by employees for each site task and operation being conducted.
- Medical surveillance requirements (employer program).

- Frequency and types of air monitoring, personnel monitoring, and environmental sampling techniques and instrumentation to be used, including methods of maintenance and calibration of monitoring and sampling equipment. This also includes establishing action levels (e.g., work stops at certain concentrations).
- Site control measures.
- Decontamination procedures.
- An emergency response plan.
- Confined-space entry procedures.
- Spill-containment program.

PROPERTIES OF CHEMICAL SUBSTANCES

In assessing hazards that may be posed by various substances personnel may be engaged in sampling, a review of the characteristics of each material to be sampled should be conducted. When conducting this hazard assessment, physical hazards associated with the site, along with the properties of chemical substances, should be documented and appropriate protective measures employed.

SITE CONTROL

Site control is intended to establish barriers isolating various contamination hazards from employees, contractors, bystanders, and off-site receptors. Site control has the following three primary goals:

- Minimize potential contamination of workers.
- Minimize the unwanted movement or spread of contaminants.
- Communicate the individual site control requirements to all of the people working at the site.

Many factors affect site control, such as these:

- Characteristics of the individual site, including physical size, weather conditions, terrain, proximity to adjacent off-site businesses or activities, and size of the contaminated area.
- Types of tasks that will be performed on the site, such as extensive excavation, on-site treatment, and sampling.
- Types of chemical contamination or potential contamination that employees may be exposed to on-site. Properties such as flammability, toxicity, and reactivity may require certain site control procedures and certain levels of decontamination.

To establish effective site control, a set of procedures must be developed that is field practical, and flexible when conditions warrant. To implement site control measures, the following tasks are necessary:

- A site map indicating areas of contamination and associated work zones. Work zones consist of exclusion (hot, or area of highest contamination), decontamination (warm, or contamination reduction), and support (cold, or equipment and supply staging) zones.

- Site entry and exit procedures (specifying where to enter and exit).
- Buddy system—should be practiced whenever possible, in particular when people are wearing protective clothing or handling hazardous chemicals.
- Site security—may include barriers, barricades, hurricane fencing, signs, or personnel stationed on-site.
- Communication system.
- Safe work practices or standard operating procedures.

DECONTAMINATION

Decontamination procedures are necessary to reduce or eliminate the spread of contamination from workers or on-site equipment to areas beyond the site control boundaries (such as the employee's clothing or protective garments, tools, home, or automobile). If contaminated equipment is allowed to leave a site, significant stretches of roadway or passing vehicles may be inadvertently contaminated. Sampling equipment that is not properly decontaminated also may contribute to erroneous sampling results on subsequent sites.

Decontamination is the process of reducing contamination on a surface or material to a safe or negligible level. Contaminants may be carried off-site by a variety of mechanisms, including track-out, air dispersion, runoff, and surface-to-surface contact. Gases, vapors, liquids, or particulates may rest on, react with, or permeate the outer surfaces of protective clothing. The following methods of decontamination are in general use:

- *Chemical decontamination* involves the use of a diluted chemical solution to detoxify, neutralize, or mobilize contaminants.
- Physical removal involves methods such as dislodging, wiping off, or other means of nonchemical removal. This method is often as simple as carefully removing your garments (*dry decontamination*).

The decontamination (decon) method selected should be based on the type of chemical, the amount, hazards present, the level of protection being employed, and resources available. A decon team should be established to be responsible for setting up the decon line and assisting personnel entering or exiting the hot zone. Key items to remember include the following:

- Everyone entering the hot zone wears the same level of protection.
- The decon area should be set up before anyone enters the hot zone.
- Decon personnel also must wear protective equipment.
- Avoid direct contact with items being decontaminated by using such methods or items as water spray, sponges, and brushes.
- Field instruments and sampling equipment carried into the hot zone must be handled carefully so as not to contaminate the instrument.
- Control runoff by collecting water in a decon pool or tarped area; do not use excess water.
- Properly containerize all protective garments and contaminated material for disposal.

QUESTIONS FOR REVIEW

1. What are some of the reasons a site safety plan might be included within or as a separate part of the sampling plan?
2. Site work zones include the hot or exclusion zone, warm or contamination-reduction zone, and the cold or support zone. What is unique about each of these work zones?
3. What must be done before selecting protective clothing and equipment?
4. Hazard assessment identifies what types of hazards?
5. What are the goals of site control? How is site control implemented?
6. What are some factors affecting site control?
7. What is the process of reducing levels of contamination to safe or negligible levels?
8. What are the two basic methods that may be selected, based upon contaminants present at the site, to decontaminate workers and equipment? Discuss each.
9. Should decontamination areas and supplies be prepared before entering an area to perform work? Why?
10. Should everyone entering a hot or exclusion zone wear the same level of protection? Why?
11. Should decontamination liquids be allowed to run off-site? If not, what should be done with them?

ACTIVITIES

1. Develop the basic elements of a site safety plan based on the sampling plan that was developed in Chapter 4.
2. Contact the state environmental or hazardous waste management agency and ask to obtain a copy of a representative site safety and health plan that the agency uses when collecting samples or a site safety and health plan for a site that has been disclosed to the public. Prepare a brief report summarizing the work zones, action levels, and specified protective clothing.

READING

Standard Methods for the Examination of Water and Wastewater, 1992. 18th ed. Environmental Protection Agency.

7

Environmental Hygiene Sampling—Gas, Vapors, and Aerosols

Dean R. Lillquist
David O. Wallace

Upon completion of this chapter, you will be able to do the following:

- Understand what an air pollutant is, the different types, exposed populations, and health effects.
- Identify the classes of airborne contaminants to better communicate the type of airborne contaminant, as well as health and safety concerns.
- Know how to express airborne contaminant concentration in either parts per million (ppm) or milligrams per cubic meter (mg/m^3), depending on which is appropriate.
- Understand why safety standards and recommendations may be very different from health standards and recommendations.
- Understand the common terms associated with health and fire safety (IDLH, LEL, UEL, flash point, etc.).
- Understand the difference between acute and chronic health effects and how this affects desired air sampling outcomes.

INTRODUCTION

Under normal conditions, ambient air (free-moving, outdoor air) contains 78% nitrogen, 21% oxygen, 9% argon, and trace quantities of numerous other chemicals found at constant levels (see Table 7–1). Other naturally occurring chemicals and human-produced contaminants (*human-produced chemicals* are referred to as *anthropogenic*) are found in the air at varying levels. These chemicals include water vapor, carbon dioxide, hydrocarbons, nitrogen oxides, carbon monoxide, ozone, sulfur oxides, and particulates. Air becomes polluted when its makeup is changed so that it is less useful or even harmful to people. Thus, things that pollute the air include both natural and anthropogenic gases, vapors, and aerosols.

The U.S. Environmental Protection Agency (EPA) is the organization responsible for regulating ambient air pollution and is generally concerned with community exposure to pollutants. The EPA recognizes that the very young, the very old, and health-compromised individuals compose part of the community. The EPA also recognizes that community exposure may occur twenty-four hours per day, seven days per week, for a lifetime. For these reasons, EPA standards are usually much lower than workplace (occupational) standards and are expressed as 24-hour or annual averages.

In addition to concerns over ambient air pollution, other environments can have air quality concerns. Significant air pollution can exist in occupational and building environments. Occupational exposure to air contaminants has been recognized for hundreds of years. These exposures are usually a direct result of chemicals associated with industrial processes or as process by-products. *Industrial hygiene* is the science and art devoted to the anticipation, recognition, evaluation, and control of occupational health hazards. Industrial hygienists traditionally have been responsible for measuring and evaluating occupational exposures. The U.S. Occupational Safety and Health Administration (OSHA) is the organization responsible for regulating worker exposure to these agents. The traditional worker is employed eight hours per day, five days per week, and is considered to be healthy and between the ages of 16 and 65. Because of these factors, OSHA standards are usually established for 8-hour exposure periods and are generally less protective (higher) than EPA standards.

A relatively new concern (since about 1970) involves the quality of air in indoor environments such as homes, offices, and public buildings. This subject is referred to as *indoor air quality* (*IAQ*). Indoor airborne contaminants have been shown to come from the building itself (carpet and paint off-gas, friable insulation, etc.), chemicals used in the building (cleaners, pesticides, office products, etc.), and chemicals introduced from the outside through building leaks and building ventilation. Both the EPA and OSHA are responsible for investigating and controlling IAQ.

TABLE 7–1
Normal constituents of air found at constant levels.

Atmospheric Constituent	Atomic/Molecular Formula	Ambient Concentration (%)
Nitrogen	N_2	78.1
Oxygen	O_2	20.9
Argon	Ar	9.3
Neon	Ne	0.02
Helium	He	0.005
Krypton	Kr	0.001

In the occupational environment, air pollutant levels may be found at such high concentrations that they pose immediate risk of explosion, fire, or acute poisoning. Sometimes air pollutants are found at lower concentrations where long-term exposure may cause chronic health effects. For these reasons, it may be necessary to evaluate the airborne concentration of pollutants or contaminants.

Reasons for air sampling can include the following:

- Determining the physical makeup of airborne contaminants (i.e., gas/vapor, aerosol/particulate, or a combination).
- Determining whether concentrations are high enough to be a safety or health concern (i.e., in excess of recognized safety or health limits, guidelines, or regulations).
- If it is a health concern, determining whether it is acute or chronic (i.e., short-term or long-term exposure and/or health effects).
- Determining whether certain areas or operations require nonroutine controls.
- Determining whether personal protective equipment (PPE) is necessary and, if so, the appropriate (adequate) type.

Two key questions to ask are (1) what contaminants are in the air and (2) why are we interested in knowing the contaminant concentration in the air. The following discussion begins by identifying and classifying chemicals based on their physical and chemical properties, followed by a discussion on how to express airborne concentrations of chemical agents. Finally, a brief look at safety issues versus health issues and why they affect air sampling methods will end the chapter. The overall goal of this chapter is to help us better understand the importance of the two previous questions and build an effective vocabulary to help us communicate with others in answering them.

CHEMICAL AND PHYSICAL PROPERTIES OF AIRBORNE CONTAMINANTS

One way contaminants can be classified is by their chemical and physical properties. This classification scheme gives us a basic understanding of how contaminants behave in air and the sampling technologies used to evaluate the airborne concentration. It is important to learn to use the correct technical terms to minimize confusion when discussing specific contaminants and sampling strategies. Airborne contaminants can be classified as either gases and vapors or aerosols. There are numerous subclasses of aerosols that also deserve further definition.

Gases and Vapors

Gas

A *gas* is a formless fluid that expands to occupy its container. A gas maintains these characteristics at room temperature. Both increased pressure and decreased temperature are required to convert a gas into a liquid or a solid.

Vapor

A *vapor* is also a formless fluid that expands to occupy its container. The difference between a gas and a vapor is that a vapor results from the evaporation of liquids (solvents) or sublimation of solids at room temperature. A vapor requires either increased pressure or decreased temperature to convert it back to a liquid or a solid.

Once a liquid or solid is in the vapor phase, the vapor behaves identically to the way a gas does. The term *vapor* gives additional insight into the origin of the airborne chemical.

Vapor Pressure

Vapor pressure is the pressure of the vapor of a substance in equilibrium with its condensed phase. It can be illustrated by the example of a closed flask with a small amount of a liquid solvent in the bottom. (See Figure 7–1). If the solvent and its vapor are allowed to come to equilibrium, the pressure exerted by the evaporated solvent on the sides of the flask is the vapor pressure.

Vapor pressure gives an indication of how fast a chemical will evaporate (liquids) or sublime (solids) into the air. Chemicals with a high vapor pressure evaporate faster than chemicals with a low vapor pressure. Table 7–2 shows the vapor pressure for methylene chloride, acetone, water, and malathion. Anyone who has watched a small amount of acetone evaporate off a surface knows how quickly this occurs. Most of us can visualize the evaporation rate of water when it is spilled on a surface. Insecticides such as malathion have been formulated to evaporate slowly so that they stay where they are applied. It is important to realize that temperature greatly affects vapor pressure, and thus evaporation rate. Higher temperatures increase the vapor pressure and the corresponding rate of evaporation.

Because all humans breathe, both airborne gases and vapors can be significant health concerns. The respiratory system offers a quick and efficient route of entry into the body. Our system's high moisture content often enables chemicals that react with water to irritate the mouth, nose, throat, and lungs. Other chemicals are absorbed into the blood through the lungs (remember that our respiratory system evolved to absorb oxygen and eliminate carbon dioxide).

Aerosols

Aerosols are solid or liquid particles (0.001–100 μm aerodynamic diameter) suspended in a gas. The term *aerodynamic diameter* is used to describe the size of an aerosol based on the way it behaves in an airstream. An aerosol's behavior is based

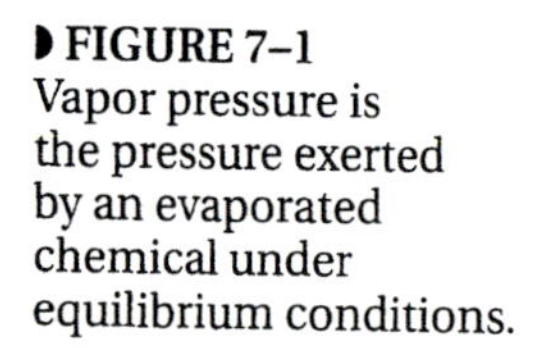

FIGURE 7–1
Vapor pressure is the pressure exerted by an evaporated chemical under equilibrium conditions.

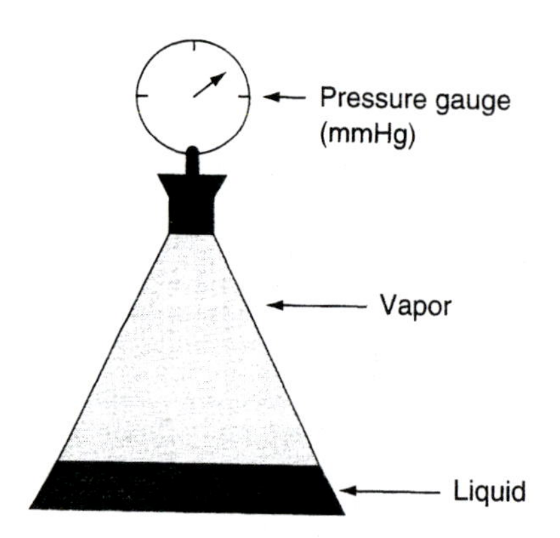

TABLE 7–2
Vapor pressures for various chemicals at normal pressure and temperature.

Chemical Name	Vapor Pressure
Methylene chloride	350 mmHg
Acetone	180 mmHg
Water	24 mmHg
Malathion	0.00004 mmHg

on its size, shape, and density. Thus, two particles may have very different shapes, sizes, and densities but because they move with the air in a similar manner, they have identical aerodynamic diameters. Many types of aerosols exist; an effort is being made to further define and standardize their subclasses.

Dust

Dusts are solid particles formed by mechanical action (abrasion) on a solid parent material. Dusts are usually irregularly shaped and larger than 0.5 μm aerodynamic diameter.

Fumes

Fumes are solid particles (<0.05 μm aerodynamic diameter) and agglomerations resulting from the recondensation of vapor. Fumes often are associated with the cooling and subsequent recondensation of melted metals or polymers (plastics). Welding fumes, for example, are common aerosol contaminants in the workplace.

Smoke

Smoke consists of solid or liquid particles (<1 μm aerodynamic diameter) resulting from the incomplete combustion of carbonaceous materials.

Mist

Mists are liquid droplet aerosols formed by the condensation of vapor or by a physical action (such as spraying, bubbling, or nebulization).

The ability of aerosols to be taken in by the respiratory system and the region of the respiratory system where these aerosols deposit are highly dependent on the aerosol's aerodynamic diameter. Particles >100 μm in aerodynamic diameter are thought to be so large that they quickly settle out of the air and are seldom inhaled into the mouth or nose. (As a comparison, the average human hair is approximately 100 μm in diameter). Only very small particles (<10 μm in aerodynamic diameter) can reach the deep lung, the area where the gas exchange between inhaled air and the blood occurs.

Size-selective aerosol samplers that collect aerosols only of a particular size are available commercially. Size-selective sampling is used when we know that the aerosol's size is of particular concern. Some standards require size-selective aerosol sampling when the risk of health hazards depends on aerosol size. For example, the EPA's particulate matter standard regulates particles only <10 μm in aerodynamic diameter (PM_{10}). OSHA's silica standard is for particles <4 μm in aerodynamic diameter.

Radio Nuclides

Additional knowledge and training are mandatory when radioactive materials are encountered. However, an introduction to relevant terms and definitions will be provided here.

Alpha Particle (Alpha Radiation, α)

Alpha particles are made up of two neutrons and two protons, giving a proton a unit charge of +2. Alpha particles are emitted from an atom's nucleus. They cause high-density ionization, transferring their energy to other matter in a very short distance. Alpha particles are stopped easily by materials such as paper, clothing, and skin. Clothing and skin are effective barriers to alpha radiation, so hazards are associated primarily with internal exposure. When alpha-emitting compounds are ingested or inhaled, or when they contaminate an open wound, internal organs and tissue are exposed directly to alpha radiation. These tissues have actively dividing cells and appear to be more susceptible to alpha radiation.

Beta Particle (Beta Radiation, β)

Beta particles are small, negatively charged particles, identical to electrons. They are emitted from an atom's nucleus and may have various energy levels. Beta particles can be both internal and external hazards because they can penetrate the skin.

Gamma Rays (Gamma Radiation, γ)

Gamma rays are electromagnetic photons of energy emitted from an atom's nucleus. They are highly penetrating and are an external radiation hazard.

EXPRESSING AIRBORNE CONCENTRATIONS

Volume-per-Volume Relationships (ppm)

Airborne concentrations of chemicals can be expressed in either a volume-per-volume (volume/volume) or weight-per-volume (weight/volume) relationship. Gas and vapor concentrations are commonly expressed as a volume/volume ratio, in parts per million (ppm), parts per billion (ppb), parts per trillion (ppt), and so on. This relationship depends on the volumes of both the gaseous contaminant and air, which correspondingly depend on pressure and temperature as described by the ideal gas law:

$$PV = nRT$$

with algebraic manipulation

$$V = \frac{nRT}{P}$$

where
P = pressure (Pa)
V = volume (L)
n = number of moles
R = gas constant (0.0821 L air/mole °K)
T = temperature (absolute)

and therefore

$$ppm = \frac{\text{Volume of gas} \times 10^6}{\text{Total volume of air}^*}$$

*At low concentrations, the small volume of contaminant present in the total volume of air is negligible.

Weight per Volume Relationships (mg/m³)

The second method for expressing airborne concentrations uses units of the weight of the chemical per total volume of air. This method can be used for expressing either gas and vapor or aerosol concentrations. Common units are grams, milligrams, or micrograms per cubic meter (g/m^3, mg/m^3, or $\mu g/m^3$). Airborne concentrations of chemicals using weight/volume ratios are independent of temperature or pressure and require no correction.

PPM and mg/m³ Equivalents

For gases and vapors, the equation for relating ppm to mg/m^3 is

$$\frac{mg}{m^3} = \frac{ppm \times MW}{24.45}$$

or

$$ppm = \frac{\frac{mg}{m^3} \times 24.45}{MW}$$

where

MW = gram molecular weight of substance.

SAFETY VERSUS HEALTH

Safety, health, and environmental concerns often overlap in real-world applications of the environmental specialists. (See Figure 7–2.) Depending on the particular organization, the amount of overlap in professional responsibility varies. However, anyone working in any one of the these areas needs some familiarization with and appreciation of the other disciplines.

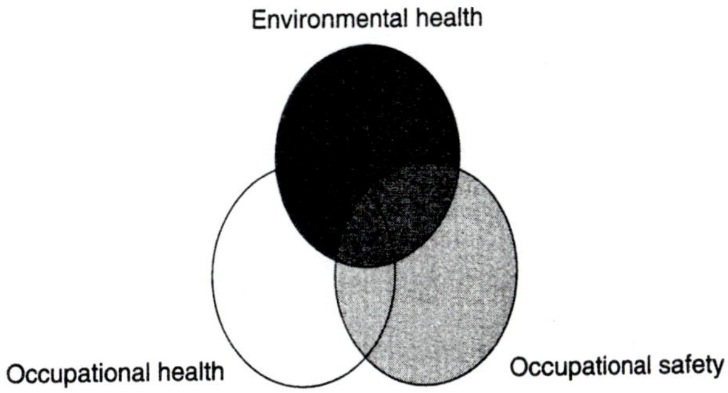

FIGURE 7–2
Diagram representing the overlap of safety, health, and environmental specialties.

Safety Concerns of Airborne Contaminants

When we discuss airborne contaminants in terms of safety, we primarily are concerned with fire or explosion hazards or chemical concentrations that are immediately dangerous to life or health (IDLH).

IDLH

The National Institute for Occupational Safety and Health (NIOSH) defines IDLH as "a condition that poses a threat of exposure to airborne contaminants when that exposure is likely to cause death or immediate or delayed permanent adverse health effects or prevent escape from such an environment." NIOSH further says that the purpose of establishing an IDLH exposure concentration is to "ensure that the worker can escape from a given contaminated environment in the event of failure of the respiratory protection equipment." Airborne concentrations that are considered IDLH have been established for more than 380 substances in the *NIOSH Pocket Guide to Chemical Hazards*. IDLH conditions will be discussed in a later section.

Explosive Atmospheres

This section is not a substitute for the very good discussion on fire safety found in the introduction to industrial safety found in textbooks. However, an understanding of the basic principles of explosive atmospheres is required to be technically aware of and to take the next step in understanding air sampling principles.

Fires and explosions are associated with high concentrations of combustible contaminants mixed with air. Airborne concentrations capable of fire and explosion are usually in percentage concentrations (1% = 10,000 ppm).

A discussion of fire or explosion must begin with an introduction to the fire tetrahedron. Figure 7–3 demonstrates the four elements required to create a fire or explosion. Fuel is any material with thermal value (i.e., can burn). Oxygen is required to combust (oxidize) the fuel. A minimum temperature (heat) is required for the oxidation to occur. Finally, a source of ignition is required to initiate the chemical reaction. If any of these four elements is missing or not in the right proportion (see the following definitions), an explosion or fire cannot occur. However, a situation in which only one element is out of balance may be very hazardous because environmental conditions can change quickly. A spark from static electricity or a small increase in or introduction of oxygen or fuel can create potentially catastrophic conditions.

FIGURE 7–3
Representation of the four elements required to create an explosion or sustain a fire.

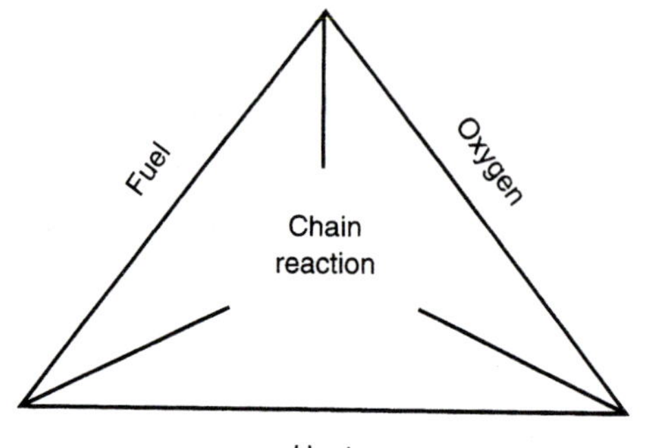

A few additional working terms must be included in our fire/explosion vocabulary. They are presented in the next sections.

Lower Explosive Limit (LEL)

Lower explosive limit is sometimes referred to as *lower flammability limit (LFL)*. LEL describes a condition in which there is sufficient oxygen and just enough fuel in the air to start a fire or explosion (see Figure 7–4) with appropriate heat and an ignition source. More fuel is required, however, for efficient combustion. To an automobile mechanic, LEL is the "leanest" fuel and air mixture that will keep a motor running. NIOSH recommends <10% LEL for conditions at which no special programs or procedures are usually necessary.

Upper Explosive Limit (UEL)

Upper explosive limit is sometimes referred to as *upper flammability limit*. UEL describes a condition in which there is too much fuel and too little oxygen to support combustion. (See Figure 7–4.) An automobile mechanic would refer to this as a "rich" mixture of fuel and air. It would cause flooding of the internal combustion engine. Although conditions above UEL technically will not cause a fire or explosion, this is still a very hazardous situation because of oxygen in the ambient air. Introduction of "fresh" air to a >UEL environment immediately produces an atmosphere that will burn or explode if there is a source of ignition.

The range between UEL and LEL is referred to as the *flammable range*.

Aerosol Explosions

Hundreds of agricultural, metal, resin, pharmaceutical, and miscellaneous compounds can produce explosive dusts. Dust explosions are similar in many respects to gas or vapor explosions, especially for particles less than 0.5 µm in diameter. Airborne dust concentrations of combustible or oxidizable solids also exhibit upper and lower explosive limits.

Flash Point

Flash point is the minimum temperature at which a liquid gives off sufficient vapor (evaporates quickly enough) to form an ignitable mixture with air at the surface of the liquid. Any combustible liquid, when heated to a temperature above its flash

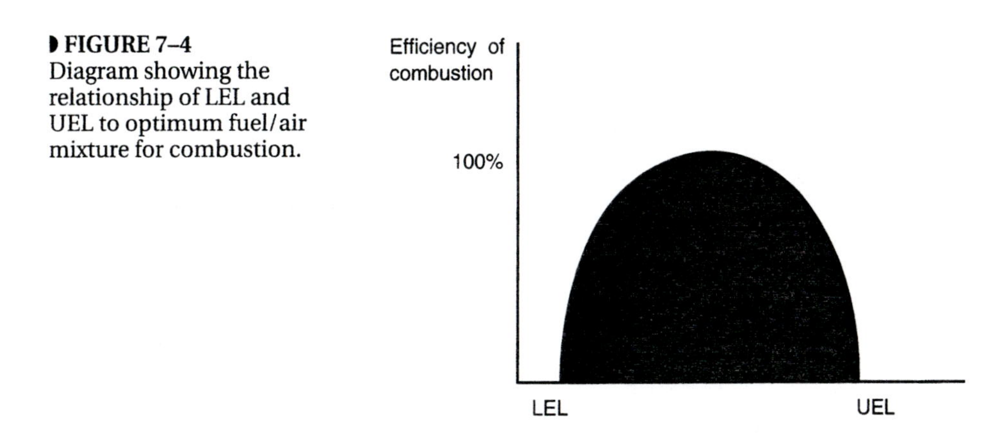

FIGURE 7–4
Diagram showing the relationship of LEL and UEL to optimum fuel/air mixture for combustion.

point, will create ignitable vapors. The lower the flash point, the greater the combustion hazard. Kerosene's flash point is about 43°C (100°F). Gasoline's flash point is about –46°C (–50°F).

Auto Ignition Temperature

Auto ignition temperature is the minimum temperature at which a flammable mixture will ignite without spark or flame.

Flammable and *combustible* refer to a classification scheme for materials based on their individual flash points. (See Table 7–3.) Using the chemical/physical characteristics defined by OSHA, Department of Transportation (DOT), and the National Fire Protection Association Standard, we can define the following terms:

Flammable

Liquids with a flash point below 37.8°C (100°F).

Combustible

Liquids with a flash point higher than 37.8°C (100°F) and lower than 93.3°C (200°F). They do not ignite as easily as flammable ones do.

Flammable and combustible liquids are further subdivided into the classes shown in Table 7–3.

When concerns center on fire, explosion, or IDLH hazards, information on airborne concentrations is required immediately. Equipment used to measure airborne concentrations must respond immediately—in real time.

Health Concerns of Airborne Contaminants

A number of health effects can result from inhaling gas, vapors, and aerosols. These health effects can be categorized as *acute* (immediate) or *chronic* (longer in duration). Some chemicals have more than one health effect and can cause both acute and chronic health problems. Acceptable atmospheric concentrations of toxic chemicals have been established by different organizations. OSHA permissible exposure levels (PELs) are legally enforceable limits that must not be exceeded. NIOSH and the American Conference of Governmental Industrial Hygienists (ACGIH) have established non-regulatory (voluntary) guidelines for allowable occupational exposure levels: NIOSH recommended exposure levels (RELs) and ACGIH threshold limit values (TLVs). These values may be the same or different from each other, as well as from OSHA's limits.

TABLE 7–3
Classifications of flammable/combustible liquids.

	Class	Flash Point	Boiling Point
Flammable Liquids	Class IA	<22.8°C (73°F)	<37.8°C (100°F)
	Class IB	<22.8°C	≥37.8°C
	Class IC	≥ 22.8° and <37.8°C	
Combustible Liquids	Class II	≥37.8°C and <60°C (140°F)	
	Class IIIA	≥60°C and <93.4°C (200°F)	
	Class IIIB	≥93.4°C	

This chapter is not intended to be a discussion of toxicology or regulations and standards. However, the toxic end point of concern and regulatory compliance issues do affect the sampling strategy employed. Therefore, the reader is encouraged to review a basic discussion of toxicologic principles and become familiar with the types of regulations (EPA and OSHA) that establish allowable airborne concentrations based on these principles.

Acute Effects

Acute health effects are immediate, ranging from irritation to life threatening. Examples of acute health effects include these:

- Irritation of mucus membranes (eyes, nose, mouth, throat, lungs, etc.).
- Central nervous system depression.
- Asphyxiation (oxygen deprivation).

Sampling instruments that have immediate or "real-time" response often are used to evaluate airborne concentrations of agents with acute effects. Occupational standards for acute toxins are expressed in terms of ceiling values—airborne concentrations that should not be exceeded, even momentarily.

Chronic Effects

Chronic health effects are diseases that take a long time to appear. These types of effects include the following:

- Internal (visceral) organ damage
  lungs
  nervous system
  liver
  kidney
  blood
- Cancer
- Dermatosis (skin problems)

Instruments used for evaluating airborne concentrations of agents when chronic health effects are of concern either can have real-time response or produce samples over an extended period of time. Integrated sampling methods give an average concentration over the sample period. Results are then compared to time-weighted average exposure regulations. Allowable exposure regulations are often expressed in terms of 8-hour time-weighted average (TWA) exposures.

SUMMARY

At the beginning of this chapter, two key questions were identified that need to be answered to develop appropriate sampling strategies: What contaminant is in the air? Why are we are interested in knowing the airborne contaminant concentration?

This chapter has explained the importance of these questions and introduced related concepts and terminology. Airborne contaminants affect the general public's health and occupations. These two populations are not necessarily the same, and different theories for protection apply. The importance of using the correct terminology was stressed to make technical communication more meaningful.

The next step is to take the information on the type of chemical contaminant, the reason it is of concern, and how quickly we need to know its airborne concentration to choose appropriate sampling methods and instrumentation.

QUESTIONS FOR REVIEW

1. Are gases and particulates resulting from volcanic eruptions considered air pollutants?
2. When a backhoe disrupts dry, compacted soil, what is the classification of the airborne aerosol generated?
3. Welding operations often create what type of aerosol?
4. If 10 μL of chlorine gas is injected into 1 L of air, what is the concentration in ppm?
5. What is the ppm equivalent for a 10 mg/m^3 concentration of toluene vapor ($MW_{toluene}$—92)?
6. Is an environment that is >UEL also IDLH? Is an environment that is IDLH also >UEL?
7. What are the four things required to create a fire/explosion?
8. Which is the more hazardous condition: <UEL or >LEL? Why?
9. What is the maximum flash point for a Class IB flammable liquid?
10. Why are environmental regulations often expressed as 24-hour averages while occupational regulations are often expressed as 8-hour averages?

ACTIVITIES

1. Using the 1994 NIOSH Pocket Guide to Chemical Hazards or a similar document, identify the chemical structure, OSHA–PEL, vapor pressure, and IDLH concentration for
   a. benzene
   b. toluene
   c. xylenes
   d. mercury
2. Light a wooden match, put it in a glass flask, and cap the flask tightly. Describe what you see and why what you see happens.
3. Put approximately 1 gram of flour on a glass evaporation dish. Using a match, attempt to light the flour. Now light another match and sprinkle a pinch of the flour over it. Describe what you see and attempt to explain the reason for the result you obtain.
4. Using a "closed cup," determine the flash point of methanol.

REFERENCES

Godish, T., 1991. *Air Quality*, 2nd ed., Chelsea, MI: Lewis Publishers, 422 pp.

Willeke, K., and P. A. Baron (eds.), 1993. *Aerosol Measurement: Principles, Techniques, and Applications.* New York, NY: Van Nostrand Reinhold, 875 pp.

National Safety Council, 1988. *Accident Prevention Manual for Industrial Operations: Engineering and Technology*, 9th ed., Chicago, IL: National Safety Council, 533 pp.

Clayton, G. D., and F. E. Clayton (eds.), 1991. *Patty's Industrial Hygiene and Toxicology, Volume I, Part B—General Principles*, 4th ed., New York, NY: John Wiley & Sons, 1091 pp.

NIOSH, OSHA, USCG, EPA, 1985. *Occupational Safety and Health Guidance Manual for Hazardous Waste Site Activities.* DHHS (NIOSH). Pub. No. 85–115.

NIOSH, 1994. *Pocket Guide to Chemical Hazards.* DHHS (NIOSH). Pub. No. 94-116.

8

Environmental Hygiene Sampling—Air-Sampling Strategies and Instruments

Dean R. Lillquist
David O. Wallace

Upon completion of this chapter, you will be able to do the following:

- Understand different air-sampling strategies.
- Know different air-sampling instruments and technologies, and how these relate to sampling strategies.
- Fully understand the need and procedures for instrument calibration.
- Know how to calculate exposure based on field sampling and laboratory results.

INTRODUCTION

An important purpose of air sampling is to identify contaminants. Often the air contaminant is a known substance, so monitoring is performed to measure the concentration in the air. Reliable measurements of airborne contaminants are useful for the following:

- Determining whether a chemical is present and at what concentration.
- Assessing the potential safety concerns of airborne contaminants.
- Assessing the potential health effects of exposure.
- Determining whether controls (including personal protective equipment—PPE) are required.
- Selecting appropriate and adequate controls (including PPE).
- Delineating areas where special controls are needed.

Initial screening measurements are often qualitative, to find whether a contaminant is present. *Quantitative sampling* determines the airborne concentration. This chapter describes various strategies for air sampling based on the type of information required. A discussion follows on various types of sampling equipment and instrumentation that are available for different sampling strategies. Next is a discussion on calibration, both instrument and airflow. The chapter concludes with an example of calculations for taking field data and laboratory data, and arriving at a time-weighted average airborne concentration.

STRATEGIES FOR SAMPLING AIR

Different air-sampling strategies involve various terms and concepts. By understanding the type of contaminant (see previous chapter) and sampling strategies, a sampling plan can be developed to help answer questions regarding airborne chemical contamination.

Personal Sampling

Personal sampling assesses the exposure of an individual. The individual wears a sampling device that accompanies him or her during work.

Breathing Zone

Breathing zone samples are a specific type of personal sample. The breathing zone is a sphere with a 1-foot radius from the nose and mouth. Air sampling in the breathing zone is representative of air inhaled by the individual.

Area Sampling

Area sampling is performed in a fixed location to assess concentrations in that area. Area sampling is used to obtain background or worst-case levels. For example, area sampling can be used to predict exposure for individuals who must enter a contaminated area.

Short-Term Sampling

Sometimes called *grab sampling, short-term sampling* is performed over a short period of time, usually less than 5 minutes. Short-term sampling is used for leak detection, to identify safety hazards (LEL) and acute health hazards (oxygen deficiency or irritation), and for comparison to short-term (ceiling limits) occupational and environmental regulations.

Long-Term Sampling

Sometimes called *integrated sampling, long-term sampling* is used to evaluate airborne concentrations over a period of time (fifteen minutes, eight hours, twenty-four hours, etc.). Long-term sampling averages the airborne concentration over the sampling period. The advantage is that the results are comparable to 8-hour occupational or 24-hour environmental standards. The disadvantage of long-term sampling is that it will not show short periods of high-level exposure.

The previous terms are used alone or in appropriate combinations to explain the sampling strategy. For example, a personal, breathing zone, integrated sample means something very different than an area grab sample.

Sampling strategies can also differ based on the desired outcome. If the worst-case exposure results are required, area measurements close to the source of contamination may be desired. For representative exposure results, numerous personal samples may be taken to evaluate the average exposure. By using and understanding these terms, one can communicate how sampling was performed.

AIR-SAMPLING INSTRUMENTS

There are two primary categories of air-sampling instruments: direct-reading and integrated sample collection devices. Each of these two categories has different methods and equipment to accomplish specific sampling strategies.

Direct-Reading Instruments

Direct-reading instruments perform sampling, analysis, and measurement while the operator receives an immediate result or readout. Direct-reading instruments are often capable of performing either personal or area samples.

Direct-reading instrumentation is one of the fastest evolving areas in the field of industrial hygiene. In the past, direct-reading instruments provided only instantaneous or grab-sample results. New electronic instruments often offer data logging, computer interfacing, and integrated sampling ability.

New technologies, brands, and models are constantly being introduced. Direct-reading instruments can become obsolete in as little as three to five years as new technology makes these instruments not only smaller, lighter, faster, and easier but also more powerful, accurate, and precise. Because of the constant evolution, this chapter will discuss instrumentation in terms of classes and in a very generic sense.

Technologies are rapidly changing, so the operators of any instrument *must* become thoroughly familiar with the operating specifics of their particular brand and model. There is *no* substitute for reading and fully understanding the specific instrument's operation and technical manual.

Electronic direct-reading instruments contain detectors that generate an electrical signal in the presence of an airborne contaminant. The signal is then displayed on a meter or digital readout.

There are a number of different ways to classify direct-reading instruments. We will begin by separating these types of instruments into compound-specific and compound-nonspecific sampling instruments. Nonspecific instruments respond to a wide range of contaminants; compound-specific devices are designed to respond only to a specific chemical.

Compound-Nonspecific Direct-Reading Instruments

Many direct-reading instrument technologies can be described as one of the following:

1. Respond to a number of different chemicals at the same time (nonselective, nonspecific).
2. Can be set and calibrated to respond to a number of different chemicals individually (selective, nonspecific).

Combustible Gas Monitors

There are many commercial brands of combustible gas monitors. Models with additional built-in detectors for specific gases are popular for confined space monitoring. Sensors for such things as oxygen, hydrogen sulfide, and carbon monoxide are commonly included on one instrument (see the section on compound-specific instruments—electrochemical cells). When this is the case, the instrument is referred to as a multiple gas meter. See Figure 8–1.

For the combustible gas meter component, the theory of operation is based on either the change in resistance of a circuit due to heat released from gas combustion (Wheatstone bridge circuit) or the change in electrical conductivity of a chemical cell in the presence of a combustible gas (electrochemical sensor). Both types require calibration and familiarity with the specific operation of the instrument.

The Wheatstone bridge circuit uses a heated, catalyst-coated element. When a flammable atmosphere is introduced, the element gets hotter and the electrical resistance of the circuit changes. This change is measured electrically and is translated into a concentration.

Many instruments produce an electrical current that may cause an explosion in a flammable atmosphere. If the instrument has been engineered to be safe to use in such environments, it is referred to as *intrinsically safe.*

Electrochemical cells adsorb flammable gases or vapors that affect the cell's electrical conductivity. The change in electrical conductivity is indicated as a concentration of combustible gas/vapor. Each chemical with flammable/combustible properties has specific LEL and UELs. A combustible gas meter's Wheatstone bridge or electrochemical cell's response is compound-dependent. Combustible gas meters are calibrated against known concentrations of specific gases. It is best to calibrate

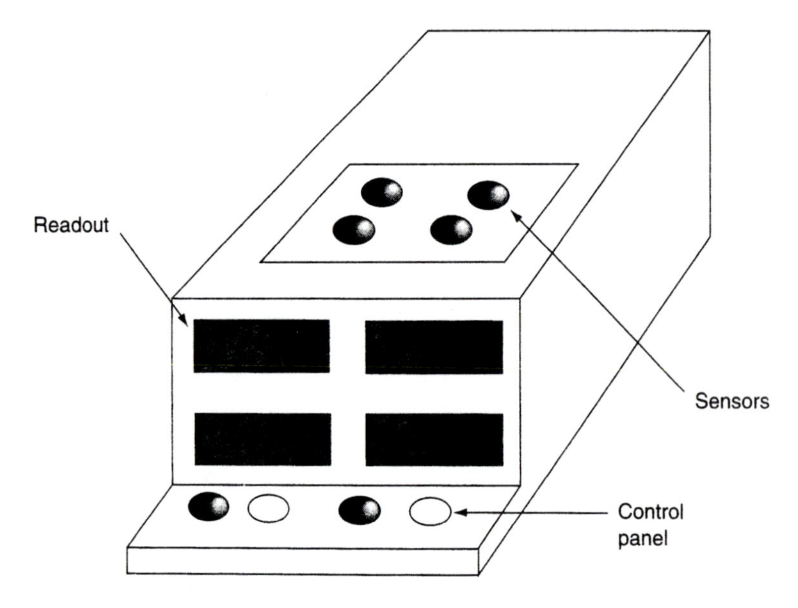

FIGURE 8–1
Portable multiple gas meter, including combustible meter.

the combustible gas meter against the chemical that will be measured. Methane or hexane is often used to calibrate combustible gas meters when measuring unknown chemicals or mixtures of combustible gases. However, under these conditions the instrument may over- or underestimate the actual concentration. Sensitivity is, thus, a function of the differences in chemical and physical properties between the calibration gas and the gas being sampled. Consult the individual operator's manual for guidelines in correlating meter readings, interfering gases, cell poisons (silicones, lead, etc.), and precautions.

Combustible gas meters usually report in terms of percent LEL (%LEL). (See Chapter 7 for a theoretical discussion of LEL.) Sometimes %LEL can be confusing because the LEL is also a percent—the percent concentration of a gas/vapor in the air. For example, a combustible gas meter is calibrated for n-hexane (LEL: 1.1% concentration of n-hexane in air) and then is taken into an n-hexane–rich environment. The meter reads 50% LEL. This means that the airborne concentration of n-hexane is 0.55%:

$$1.1\% \text{ (n-hexane's LEL)} \times 0.5 \text{ (instrument reading)} = .55\% \text{ n-hexane in air}$$

Instruments that alarm at 10–20% of LEL (in our example, 0.11–0.22% concentration of n-hexane in air) have a significant safety factor built into them.

Flame Ionization Detectors

Flame ionization detectors (FIDs) are used for monitoring compounds that ionize in the presence of a flame (often an oxy-hydrogen flame). FIDs have a small flame inside (to burn the sample) and an ion collector. As ions are collected, an electric signal is generated, which is displayed as a concentration. The FID is a very sensitive detector and can measure down to nanogram/m^3 concentrations in the air (1,000 nanograms = 1 microgram). FID is a nonspecific detection mechanism that responds to many organic-based compounds. To measure specific compounds or classes of compounds, a response factor must be determined. As the number of carbon-hydrogen bonds in the contaminant decrease, the sensitivity also decreases. FIDs do not respond to inorganic compounds.

Photo Ionization Detectors

Photo ionization detectors (PIDs) are used for monitoring compounds that ionize in the presence of ultraviolet (UV) light. Ions are collected and the electric signal is displayed as a concentration. Several different lamps are often available, each emitting a specific wavelength of UV energy. By interchanging lamps, some chemical selectivity is achieved. Stable air constituents (oxygen, nitrogen) or compounds with a higher ionization potential than the lamp's UV energy output (methane, etc.) are not ionized. Because water vapor is ionized, changes in absolute humidity may affect instrument readings. PIDs are usually calibrated against known concentrations of butadiene, and readings are reported relative to this compound.

Infrared

Many gases and vapors absorb certain characteristic frequencies of infrared (IR) radiation. Specific wavelengths of IR energy can be used to identify and quantify airborne contaminants. A broadband infrared source is used with an IR-absorbing filter to generate discrete wavelengths of IR energy. As IR energy passes through the air, contaminants may absorb IR energy. This causes a reduction in IR energy at the detector and is electrically displayed as an airborne concentration. When a number of IR-absorbing contaminants are present, separation and specific analysis may be possible due to the overlap of the individual chemical's absorbing spectra.

Aerosol Monitors

Several types of aerosol monitors are available to directly measure airborne aerosol concentrations. Most monitors are based on the fact that aerosols absorb and/or reflect light. The light-scattering properties of aerosols that make them sensitive to this method include size, shape, and refractive index. Aerosol-laden air is introduced into a chamber with a source of light. The light that reflects from the aerosols is detected and an electrical signal is generated. Instruments should be calibrated with particulates of a size and refractive index similar to those being measured.

This technology is now widely used on portable respirator fit-testing instruments. These instruments use a vapor-saturated chamber to enlarge extremely small particles by nucleicondensation. This allows very small particles to grow large enough to reflect sufficient light.

A second type of particulate-monitoring device is based on piezoelectric crystal technology. As aerosols are separated from an airstream and impact out on a piezoelectric crystal, the frequency of the crystal's oscillations is changed. This is translated into an electrical readout.

Chemical-Specific Direct-Reading Instruments

Detector (Colorimetric Indicator) Tubes

Detector tubes are a simple and inexpensive method of obtaining quick results on airborne concentrations of gases and vapors. They are truly indispensable tools. To use a detector tube, simply break the two ends and insert the tube into the manufacturer's pump. See Figure 8–2.

Following the manufacturer's instructions, the correct number of pump strokes (volume) is pulled through the tube. The length of stain (color change) on the tube correlates to a calibration scale. The scale is an indication of the airborne concentration.

FIGURE 8–2
Colorimetric indicator tubes and sampling pump.

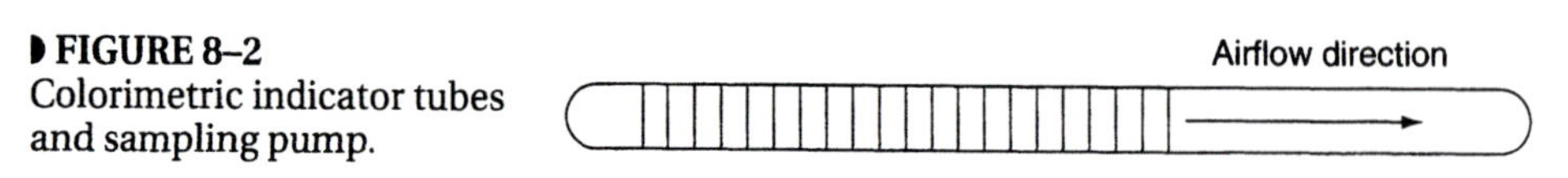

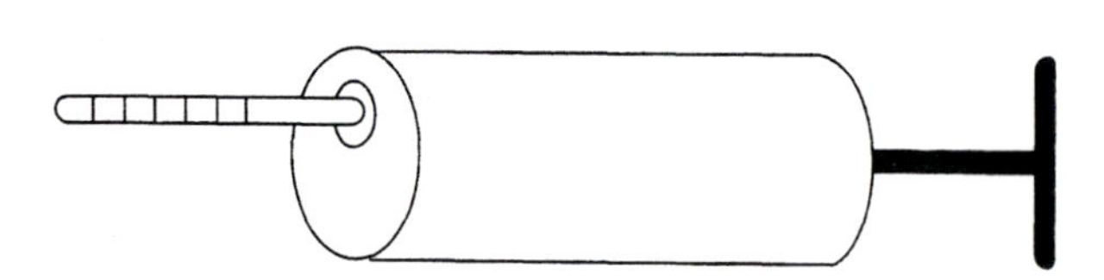

A second type of colorimetric indicator tubes does not use pumps. Passive dosimeters rely on the diffusion of chemicals across a concentration gradient (see the discussion on passive dosimeters). For this reason, a minimum airflow is required across the tube opening (usually >50 feet per minute). The time of sampling is recorded, the length of stain observed, and these two results are combined to give an indication of the average concentration over the sampling period.

Passive dosimeter badges also are available. A reactive patch on the badge changes color when it is exposed to specific chemicals in the air. The degree of color change is used to estimate the average contaminant concentration.

The primary advantage of colorimetric dosimeter technology is that it is relatively inexpensive and provides immediate results at the end of the sampling period.

Electrochemical Sensors

Electrochemical sensors rely on the principles of conductivity, potentiometry, coulometry, or ionization. They all are based on the generation of electrical signals from chemical reactions. These sensors usually rely on the passive diffusion of a contaminant across a semipermeable membrane into the reaction cell.

Instruments that use electrochemical cells can be small enough to fit into a shirt pocket. Many are equipped with integrating electronics that allow for alarms, time-weighted average (integrated) exposure estimates, and event recording such as the peak concentration and the time it occurred.

Long-Term (Integrated) Air Sampling

Integrated air sampling involves removing a contaminant from an airstream by capturing a gas, vapor, or aerosol with a collection matrix. Contaminant capture is accomplished via absorption into a collection fluid, adsorption onto a collection matrix, or filtration onto a collection matrix. Some type of analysis is required after the sample is collected. This analysis can include microscopic analysis, gravimetric (weight) analysis, or chemical analysis in a laboratory.

The sampling method you use should be tested and confirmed reliable (quality assurance/quality control). The goal is to be able to compare your sample directly to a standard or to samples taken by someone else at some other time. Standard methods involve the side-by-side development of both sampling and analysis methods. *Standard methods* are recognized, reliable methods developed by various organizations. Nonstandard methods have no proven reliability and should be avoided.

EPA, OSHA, and NIOSH have well-recognized and accepted sampling and analytical methods for industrial and environmental hygienists. In the past, these methods were found in documents, technical manuals, or within regulations. Recently, however, these organizations have published these methods on electronic media. You should become familiar with these references in hard copy or electronic versions.

In many respects standard methods are "cookbooks." Standard methods will take you step-by-step through the sampling procedure. They offer the ability to have good reproduction and comparison of sampling results.

Collection Bags

One strategy for collecting air samples is to meter air into a collection bag at a known flow rate for a known period of time. See Figure 8–3.

Many commercially available bags are designed specifically for sampling. They are made of selected materials to minimize the loss of chemicals through the bag's matrix. For example, this sampling method is recommended by NIOSH for the assessment of benzene by portable gas chromatograph (GC). The method describes metering air at 0.02–5 L/min. into a Tedlar bag until it is <80% full (at 0.02 L/min., an 8-hour sample would require a 12-L sample bag). Once the sample is taken, the air is analyzed directly by GC with a PID. The method has a stated overall accuracy of ±27% down to 0.02 ppm.

Absorption

Gases and vapors are removed from the airstream by *absorption* into a liquid medium. The liquid medium is often a specially formulated solvent or reagent. The most common application of absorption uses a fritted bubbler. See Figure 8–4.

Many tiny bubbles are formed as air is forced by suction through a porous glass frit (fritted glass). The small bubbles increase the surface area of the air in contact with the liquid collection media, increasing absorption efficiency. Removal is a function of surface area and contact time.

The impinger's liquid medium is sent to a laboratory for analysis. One NIOSH method for the sampling of formaldehyde uses impingers filled with sodium bisulfite solution.

Adsorption

Adsorption relies on physically attaching the unchanged gas/vapor onto the surface of a collection substrate. This method commonly uses granular sorbents such as activated charcoal and silica gel. See Table 8–1.

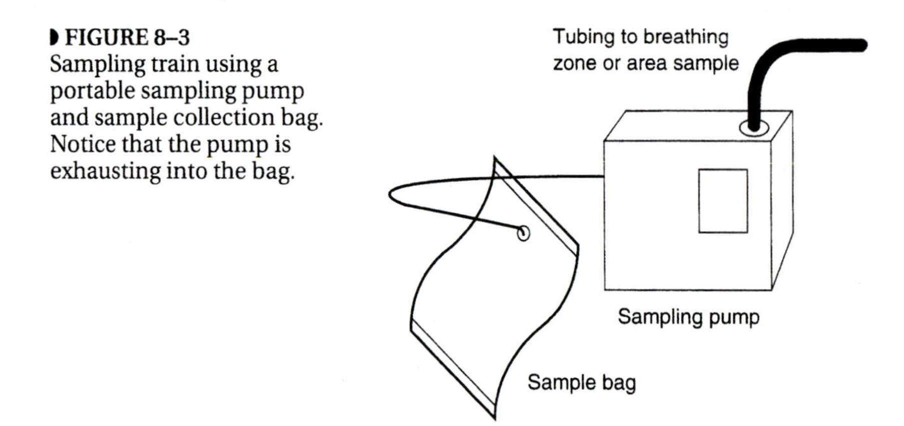

FIGURE 8–3
Sampling train using a portable sampling pump and sample collection bag. Notice that the pump is exhausting into the bag.

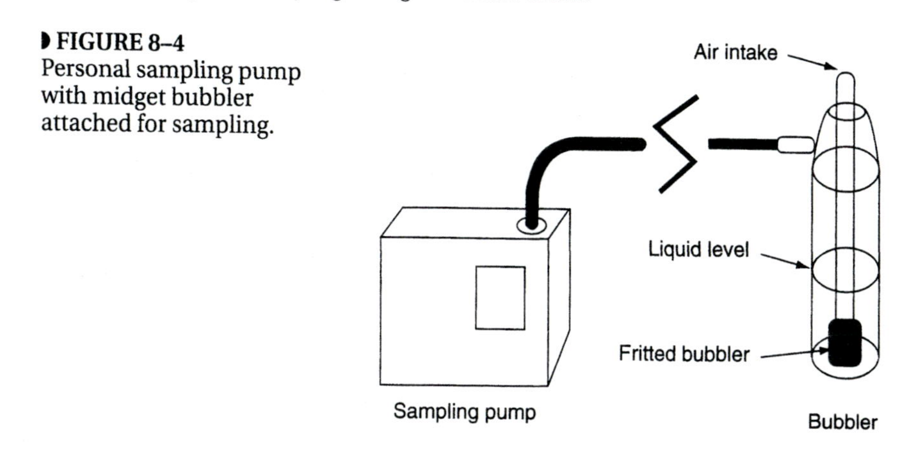

FIGURE 8–4
Personal sampling pump with midget bubbler attached for sampling.

Active Adsorption Sampling

Active adsorption sampling is very common in industrial hygiene. The adsorbing medium is contained in a glass tube. The correct tube size and medium are chosen according to standard methods. Both ends of the factory-sealed sorbent tube are broken and the tube is placed in the sampling train. The "sampling train" includes a personal sampling pump, plastic tubing, and a glass sorbent tube. For sorbent-tube sampling, pumps are calibrated for low flow (usually <100 mL/minute). See Figure 8–5.

Air is pulled at a constant volume through the adsorbing medium for a prescribed period of time. After sampling, the tube is resealed and sent to the laboratory for analysis.

Passive Adsorption Sampling

Numerous commercially available passive monitoring devices (dosimeters) have become available. The collection (or sampling) rate for passive samplers is a function

TABLE 8–1
Common sorbent materials and compound classes.

Sorbent Material	Compounds
Activated charcoal	Hydrocarbons Halogenated hydrocarbons Esters Ethers Alcohols Ketones Glycol ethers
Silica gel	Amines Phenols Nitrocompounds Aldehydes Anhydrides
Coated sorbents	Numerous compounds

FIGURE 8–5
A low-flow sampling pump with solid adsorbing tube as the collection medium.

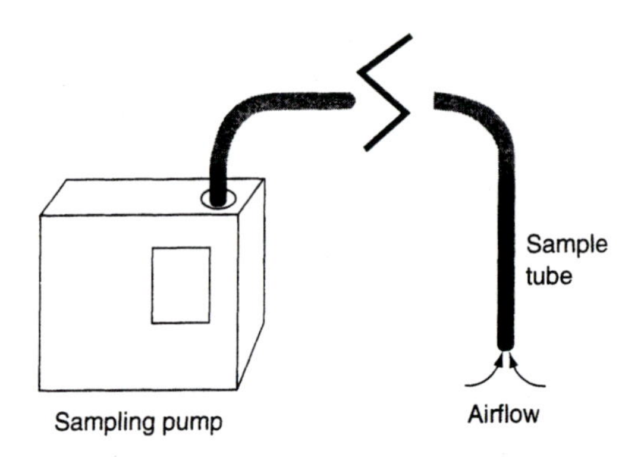

of specific chemical diffusion and permeation rates, and the geometry of the specific device. See Figure 8–6.

Passive samplers are sealed at the manufacturer. When the seal is broken, the adsorption medium is exposed to air and sampling begins. A minimum air velocity at the cover (usually >50 feet per minute) is required. Sampling stops when the device is resealed. The length of time samplers are exposed to air is recorded, and they are sent to a laboratory for analysis and results. The advantage of integrated passive samplers is that they are inexpensive and don't require expensive sampling pumps. They have proven to have good accuracy and precision. Passive samplers for many chemicals are becoming available.

Direct-reading passive samplers (dosimeters) that give length of stain or color change results are also available. See the section for colorimetric indicator tubes for a brief discussion of this technology.

Filtration

Airborne aerosol evaluation usually involves filtration and subsequent gravimetric or chemical analysis. Air to be sampled is drawn past a filter medium using a personal sampling pump. Pumps usually operate at a flow rate of approximately 2 liters per minute. See Figure 8–7. The filter medium (membrane) is held in place with a plastic cassette.

FIGURE 8–6
A passive integrated sampler.

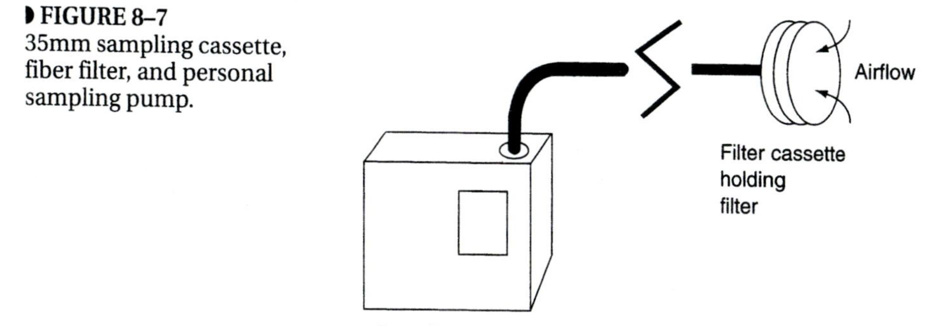

FIGURE 8–7
35mm sampling cassette, fiber filter, and personal sampling pump.

Particles are collected on the fiber membrane of the filter. There are many different filter materials (mixed cellulose, glass fiber, polyvinyl chloride, Teflon, etc.) and filter pore sizes. Filter selection depends on the material to be sampled and the analysis that will be performed. Using the *gravimetric sampling* procedure, accurate presampling and postsampling weighing of the filter is performed. Filters are usually "conditioned" at controlled temperature and humidity prior to both pre- and postweighing to minimize the effects of atmospheric moisture collecting on the filter matrix and affecting results. Once again, you are encouraged to become familiar with EPA, OSHA, and NIOSH methods calling for specific types of sampling methods and materials.

Total dust, or *total suspended particulate* (*TSP*), sampling uses the total weight of dust collected in a known volume of air. This method will provide results of airborne concentrations of TSP expressed as micrograms per cubic meter of air (g/m^3) or milligrams per cubic meter of air (mg/m^3).

Sometimes we are interested only in smaller particles, the type that can penetrate deep into the lungs and respiratory system. When this sampling is required, size-selective particulate sampling methods are used. Collection devices have been designed to remove or "preseparate" particles of a certain aerodynamic diameter (size) and larger. In turn, only the smaller particles pass by the preseparator to be collected on the filter matrix. Preseparation technology uses impaction or cyclone separation to achieve the removal of large particles.

There are a number of preseparators designed for specific environmental and occupational standards. Preseparators with a 50% collection efficiency of 10 μm in aerodynamic diameter are used by the EPA for ambient air pollution work. This sampler is often referred to as a *PM10* sampler. Occupational standards have been created for inspirable (<100 μm aerodynamic diameter), thoracic (<10 μm aerodynamic diameter), and respirable (<4 μm aerodynamic diameter) particles. Consult standards for size-selective sampling requirements and manufacturers of preseparators with the proper aerosol size-separation efficiencies.

AIRBORNE RADIATION

Radioactive contaminants are usually considered separately from chemical contaminants. This is partly because the separate science of health physics is concerned specifically with radio nuclides and partly because there is a separate regulatory agency responsible for controlling these materials. A brief discussion follows that identifies and describes the three most common field instruments used to monitor radiation.

The most common type of field survey instruments include ionization chambers, proportional counters, and Geiger-Mueller counters. An internal electric field first is established across a gas-filled chamber inside the instrument. Incident radiation ionizes gas inside these chambers. Ions are collected, and the current produced is measured and translated into an electrical signal.

Ionization Chambers

Portable ionization chambers are common survey meters. When they are used for alpha or beta particle detection, they must have a very thin window at the entry point of the chamber. This allows alpha and beta particles to penetrate into the chamber. The number of ion pairs formed in the chamber depends on the gas density. Pressurized ionization chambers increase sensitivity and are often used to measure gamma fields.

Proportional Counters

Proportional counters are very similar to ionization chambers. By increasing the voltage of the electric field in the chamber, each ion pair gains sufficient energy to ionize other gas molecules, resulting in secondary ion formation. This in turn amplifies the current and makes the proportional counters more sensitive.

Geiger-Mueller Counters

By increasing the voltage of the electric field in the chamber even higher, a single ionization will result in secondary ionization of all gas molecules in the chamber. Any ionization will result in a signal of the same magnitude. The chamber is an all-or-nothing type of monitor and counts the number of ionizing events that take place. Geiger-Mueller (G-M) counters use quenching gases or electronic quenching to stop the electrical discharge associated with each ionization. G-M tubes are useful for detecting gamma radiation that may cause only a single ionization in a gas volume.

CALIBRATION

Calibration is required for most instruments. It is used to ensure that the instrument is operating within acceptable and known parameters. Calibration is required because specific electrical components of an instrument can drift, stray, or change over time. Instrument calibration is recommended before and after sampling.

CALIBRATION OF DIRECT-READING INSTRUMENTS

Direct-reading instruments require calibration against known concentrations to ensure that they are working within specifications. Many calibration gases are commercially available, often in small compressed gas cylinders. Cylinders may contain mixtures of several compounds for multiple gas meter calibration. All that is required is to attach the correct valve and meter the gas into the instrument's sample port.

Calibration standards can be "built" by injecting a known concentration of chemical into a known volume of air. This technique requires an appropriate container (sample bag), a consistent method of metering the known volume of air into the bag, and an appropriate micro-syringe for injecting chemicals (gases or volatile solvents) into the bag. The equations for building calibration standards from gases or liquids are as follows:

Gas Calibration Standards

$$C_{(ppm)} = \frac{\text{Injection sample volume}_{(\mu L)}}{\text{Container volume}_{(L)}}$$

or with algebraic manipulation:

$$\text{Injection sample volume}_{(\mu L)} = \frac{C_{(ppm)}}{\text{Container volume}_{(L)}}$$

Liquids Calibration Standards

$$C_{(ppm)} = \frac{\rho V_1}{M} \times \frac{10^3}{V_2} \times 24.45 \times \frac{760}{P} \times \frac{T}{298}$$

or with algebraic manipulation:

$$V_1 = \frac{C\ V_2}{(10^3)\ \rho} \times \frac{M}{24.45} \times \frac{298}{T} \times \frac{P}{760}$$

where

C = desired concentration in ppm
ρ = density of liquid (gm/mL)
V_1 = sample liquid volume (μL)
V_2 = volume of container (L)
M = molecular weight of chemical
P = atmospheric pressure (mmHg or torr)
T = temperature (K [273 + °C])

Comparing calibration samples requires equipment and experience.

Although a multiple-point calibration is desirable, at minimum a two-point calibration is required. See Figure 8–8. The first calibration point is usually "fresh air," or zero-concentration air (nitrogen may be used). The instrument should first be *zeroed*. Next a calibration gas with a known concentration of chemical is introduced. The *span* on the instrument is then adjusted so that the instrument reads correctly.

Airflow Calibration

For active sampling using absorption, adsorption, or filtration, an accurate measurement of the sampling pump's airflow rate is required. Flow-rate data are used to calculate the total volume of air sampled. Instruments used to measure airflow rate can be divided into primary and secondary standards.

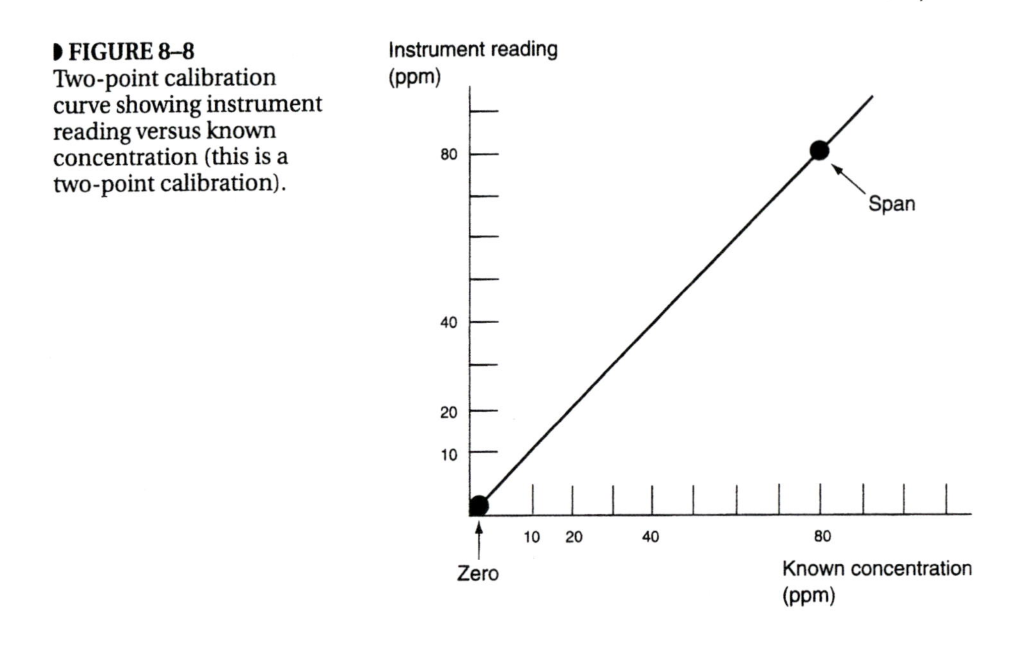

FIGURE 8–8
Two-point calibration curve showing instrument reading versus known concentration (this is a two-point calibration).

Primary Airflow Standards

Primary airflow standards generally rely on volume displacement. The advantages of primary standards are that they are not affected by temperature and pressure differences, and that they never require internal calibration. The soap bubble meter has long been used for industrial hygiene measurements. Other primary standards are generally too cumbersome for field use.

Frictionless Piston Meter or Soap Bubble Meter Soap bubble airflow calibration meters have wide use and applications for measuring airflows from 0.001–10 L/min. See Figure 8–9. A soap bubble is created at the base of an inverted graduated cylinder. As gas (air) is pulled through the cylinder by the sampling pump, an effectively frictionless piston is created as the soap bubble travels up the cylinder. A flow rate for the pump is measured by timing the bubble as it moves through a known volume of air.

Note that Figure 8–9 shows a complete sampling train (pump and sampling medium) used during the pump calibration. Using a representative collecting medium and complete sampling train helps to control any airflow resistance from the sampling medium.

Many electronic soap bubble meters are currently on the market. Although the theory of their operation is based on the primary standards, they periodically have to be sent to the manufacturer for maintenance and calibration of their electronic components. For this reason, they are sometimes called *intermediate standards.*

Secondary Airflow Standards

Secondary standards are reference instruments that trace their calibration to primary standards. However, many secondary standards are capable of exceptional accuracy and require minimal care. For these reasons, secondary standards are commonly used in field situations.

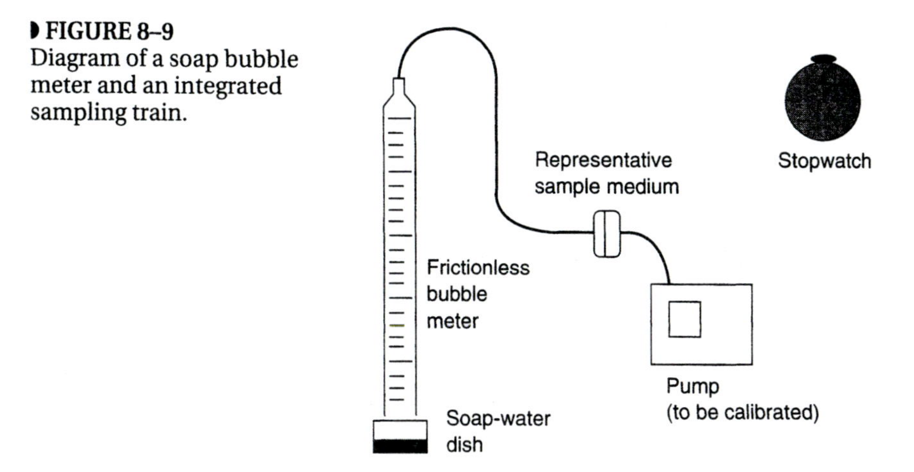

FIGURE 8–9
Diagram of a soap bubble meter and an integrated sampling train.

Rotameters Many secondary standards are cumbersome and difficult to use in the field. Rotameters, however, are very common airflow-measuring devices. Rotameters consist of a free-moving float inside a graduated (cross-hatched and often numbered) vertically tapered tube. See Figure 8–10. The tube is tapered so that the cross-sectional area is larger at the top. Air flowing through the tube will have a higher velocity in the narrow lower portion of the tube. Air flowing up the tube causes the float to rise until the air pressure on the float is just sufficient to support it. The float becomes stable—riding a current of air. The height of the float indicates the flow rate. Most rotameters use a spherical float, which is read at its widest point (at the center of the ball).

A rotameter calibration chart must be prepared with the aid of a primary standard. See Figure 8–11.

Precision rotameters are preferred for accuracy. These usually are made of glass with a 150-mm (6-inch) tube length. Floats are often stainless steel. Precision rotameters may be supplied with a calibration curve that is accurate at normal temperature

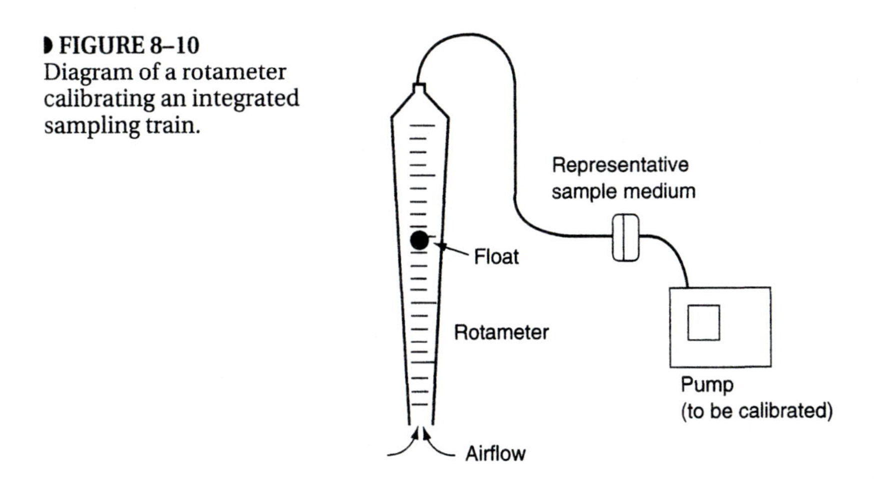

FIGURE 8–10
Diagram of a rotameter calibrating an integrated sampling train.

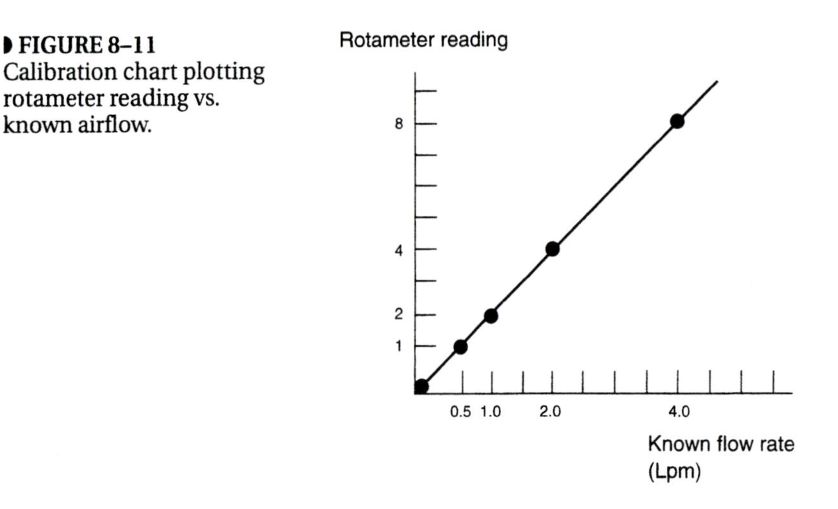

FIGURE 8–11 Calibration chart plotting rotameter reading vs. known airflow.

and pressure. A rotameter must be recalibrated or the calibration curve must be corrected if it is used at other than normal temperature and pressure conditions (or at a temperature and a pressure that are different from when it was calibrated). The equation for rotameter corrections is as follows:

$$Q_{ind} = Q_{cal}\sqrt{\frac{T_{amb}}{T_{cal}} \times \frac{P_{cal}}{P_{amb}}}$$

where

Q_{ind} = rotameter reading at conditions
Q_{cal} = flow rate at calibration conditions
(normal conditions—760 torr and 293°K [273 + 20°C]
T_{amb} = ambient temperature (°K [273 + °C])
T_{cal} = temperature at calibration conditions
P_{cal} = pressure at calibration conditions mmHg (torr)
P_{amp} = ambient pressure (torr)

Air-Sampling Calculations

This section looks at the calculations required for air sampling. Every person who conducts air sampling or equates the results must be able to perform these calculations.

Sampling Pump Flow Rate

Pumps should be calibrated before and after sampling. For example, based on three tests using a soap bubble meter, a pump takes an average of 29.4 seconds to displace 1 liter of air. The flow rate for the pump is as follows:

$$\frac{1\ L}{29.3\text{sec}} \times \frac{60\ \text{sec}}{1\ \text{min}} = 2.05\ \text{Lpm}$$

If pre- and postpump flow rates vary by more than 10–15%, the sampling results may be invalid. Smaller variations are acceptable; however, a decision is required on the desired results. If a worst-case concentration exposure estimate is desired, use the lower of the pre- and postcalibrations. Otherwise, use the average of the pre and post. In our example, the pump's precalibrated flow rate is 2.05 Lpm and its post-calibrated flow rate is 1.99 Lpm.

$$\text{Average flow rate} = \frac{2.05 + 1.99}{2} = 2.02 \text{ Lpm}$$

Air Volume Sampled

For this calculation, simply multiply the pump's flow rate by the sample duration. First, calculate the length of time the sampling occurred (in minutes). Then calculate the total volume in liters. Finally, convert liters to cubic meters.

Our sampling time started at 8 AM (08:00) and ended at 4 PM (16:00), so our total sampling time (in minutes) is

$$4 \text{ hours} \times 60 \text{ min./hr.} = \mathbf{480\ minutes}$$

To calculate the total volume sampled, take the average flow rate (2.02 Lpm) and multiply it by total sample time, as follows

$$2.02 \text{ Lpm} \times 480 \text{ minutes} = \mathbf{969\ L}$$

Finally, convert this to cubic meters

$$969 \text{ L} \times 1 \text{ m}^3/1000\text{L} = \mathbf{0.97\ m^3}$$

Time-Weighted Average Contaminant Concentration

Average concentration is found by dividing the amount of chemical recovered from the sample medium by the total volume of air sampled. The amount of chemical recovered from the sample medium is provided by the laboratory (see the chapter on industrial hygiene chemistry—laboratory analysis). The laboratory will provide results based on the weight of the chemical recovered.

For our example, laboratory results indicate that **3.20 mg** of chemical were recovered from the sample. To figure the time-weighted average concentration, take

$$\frac{3.20 \text{ mg}}{0.97\text{m}^3} = \mathbf{3.3\ mg/m^3}$$

Example of Pump Calibration Log

There are many ways to record integrated sampling data. A sample log sheet should be constructed that will have space to record any information deemed relevant to the sampling. An example of a generic sampling sheet is presented in Table 8–2. This type of sheet can be stored in a three-ring binder for data archiving or modified into a spreadsheet for computer storage.

TABLE 8–2
Example of field sampling sheet.

Date (1)	1/15/96 (2)	Time (3)	7:30 a.m. (4)	Temp (5)	22°C (6)	(7)	Press (8)	760 mmHg (9)	Signed (10)	K.L.T.

Pump #	Pre-Cal Flow Rate (Lpm)	Start Time	Stop Time	Run Time (4)–(3) (minutes)	Post-Cal Flow Rate (Lpm)	Average Flow Rate [(2) + (6)]/2 (Lpm)	Volume of Air Sampled (5) × (7) × (0.001) (m^3)	Laboratory Results (mg)	Average Concentration (9)/(8) mg/m^3
A-123	2.05	08:00	16:00	480	1.99	2.02	0.97	3.20	3.3

SUMMARY

The goal of air sampling is to determine the amount of vapor, gas, or aerosol that is present in the air under the duration and conditions at which the air sampling was performed. This information is used to ensure regulatory compliance or to assess the hazard of a given environment. In some cases immediate information is needed (for instance, before entering a confined space). At other times, 8- to 24-hour time-weighted average exposure estimates (to compare to EPA or OSHA regulations) are required. Understanding the reason for sampling and the operation, calibration, and data interpretation are crucial to ensure that the results are as accurate, precise, and appropriate as possible. Understanding instrument limitations also is important to avoid overestimating the accuracy and precision of your results.

This chapter provided an introduction to the types of instruments available to sample for airborne chemical contaminants. It was by no means an exhaustive discussion, because this is a dynamic and growing area of environmental/industrial hygiene. There is no substitute for reading and understanding the technical manuals provided with every air-sampling instrument.

QUESTIONS FOR REVIEW

1. What are the differences (in terms of how, where, and why) between (a) personal versus area sampling and (b) grab versus integrated?
2. What are the uses for and advantages to direct-reading air-sampling instruments?
3. What are two scenarios where a multiple gas meter could be used?
4. What is the difference between accuracy and precision?
5. What are several reasons that colorimetric indicator tubes are so versatile?
6. What are the three principal parts of an integrated sampling train? Draw and label one.
7. You are asked to sample for airborne toluene (MW—92) in a workshop. Based on the following results, what is the average airborne concentration?

   Precal pump flow rate—97 mL/min
   Sampling start time—8 AM
   Sampling stop time—12 noon
   Postcal pump flow rate—94 mL/min

   Laboratory results—2.00 mg of toluene was recovered.

   Assuming NTP conditions, calculate the concentration in ppm (see Chapter 7).

8. How much chlorine gas would have to be injected into 10 L of air to create a known concentration of 100 ppm?
9. How much liquid toluene (MW—92, density—0.87 gm/mL) would have to be injected into 10 L of air to create a known concentration of 100 ppm (assume NTP conditions)?
10. What is the difference between a primary and secondary standard?

ACTIVITIES

1. Select the appropriate colorimetric indicator tube and the appropriate pump for carbon dioxide (CO2). Review the manufacturer's instructions on the operation as well as the recommended number of pump strokes. Sample the ambient air, and read the length of stain. Document the entire process, including your results.
2. Construct an integrated sampling device as depicted in Figure 8–9. (*Note:* You may use a conventional bubble meter or an electronic device.) Allow the pump to run for five minutes before calibrating. Perform the flow rate calibration three times, record the results of each run in liters per minute (Lpm), and calculate the average flow rate.
3. Construct an integrated sampling device as depicted in Figure 8–9. (*Note:* You may use a conventional bubble meter or an electronic device.) Allow the pump to run for five minutes before calibrating Perform the flow rate calibration three times, record the results of each run in liters per minute (Lpm), and calculate the average flow rate. Then remove the tubing at the bubble meter and attach it to a precision rotameter (Figure 8–10). Record the rotameter reading. Adjust the flow rate on the pump and repeat this for three different flow rates. Finally, plot a calibration curve for the rotameter (Figure 8–11).
4. Using any real-time instrument available, obtain canisters of both zero (inert) gas and an appropriate span gas. Familiarize yourself with the instrument by reading the manufacturer-supplied manual. Identify the zero and span adjustments. Following the manufacturer's instructions, zero and span the instrument. Measure the unknown concentration in the gas bag supplied by your instructor.

REFERENCES

Eller, P. M., and M. E. Cassinelli (eds.), 1994. *NIOSH Manual of Analytical Methods*, 4th ed., Cincinnati, OH: U.S.D.H.H.S., CDC, NIOSH.

Cohen, B. S., and S. V. Hering (eds.), 1995. *Air Sampling Instruments for Evaluation of Atmospheric Contaminants*, 8th ed. Cincinnati, OH: American Conference of Governmental Industrial Hygienists, Inc., 651 pp.

9

Groundwater Monitoring Concepts

Michael A. Williams

Upon completion of this chapter, you will be able to do the following:

- Know the primary guidance documents that can be referenced to construct a monitoring program or interpret the results of monitoring activities.
- Describe the necessary planning activities and assessment strategies that will result in the collection of meaningful monitoring data.
- Understand drilling and monitoring well construction techniques to assist you with the interpretation of data collected during subsequent monitoring activities.
- Know the importance of following standard protocol during groundwater sampling to aid you in minimizing sources of variability.
- Become aware of the standard methods and statistical controls that a hydrogeologist uses to present groundwater monitoring data.

INTRODUCTION

Because human health and the environment are intimately linked to the quality of water resources, the effective use and protection of groundwater supplies have become major concerns. In the United States, our dependence on groundwater underscores the importance of this vital, yet sometimes neglected, economic resource. During recent decades, hydrogeologists have been dedicated to efficiently utilizing limited groundwater resources, detecting contamination, and restoring contaminated aquifers to a usable condition. *Groundwater monitoring,* the sampling and measurement of specific groundwater characteristics, is an important tool for the hydrogeologist in assessing water quality of a geologic horizon that is capable of producing significant amounts of water (an aquifer).

As you begin to understand the processes that affect our groundwater resources, you should soon realize that groundwater is not under static conditions; rather, the geochemical, biological, and physical transport phenomena that exist within an aquifer result in changing groundwater quality in response to complex interactions with the aquifer matrix, contaminants, and geochemical processes. Monitoring programs are implemented to track the dynamic and evolving chemical characteristics of groundwater with respect to temporal and spatial relationships.

The intent of this chapter is to allow you to become familiar with the basic elements of a groundwater sampling plan, recognize the design considerations for constructing a groundwater monitoring well, and understand how monitoring data are used in hydrogeologic assessments. It is not to provide detailed instruction regarding actual soil or groundwater sampling techniques. Through the examination of these basic issues, you will develop an understanding of the most fundamental tool used in environmental hydrogeology—groundwater monitoring data.

LITERATURE OVERVIEW

Groundwater monitoring techniques, principles, and practices have been discussed exhaustively in a wide array of references. The primary sources for standardized monitoring protocol are presented in publications of the U.S. Environmental Protection Agency (EPA), the U.S. Geological Survey (USGS), and professional associations such as the American Society for Testing and Materials (ASTM) and the National Water Well Association (NWWA).

Two EPA documents serve to broadly define the standard of care for groundwater monitoring protocol in the environmental industry. First, the *Ground-Water Monitoring Technical Enforcement Guidance Document (TEGD)* of the Resource Conservation and Recovery Act (RCRA) describes in detail the essential components of an appropriate groundwater monitoring program. It is intended to be used by environmental professionals, regulatory agencies, and attorneys to evaluate the adequacy of a groundwater monitoring program. The *TEGD* focuses on aspects of site characterization; monitoring well placement, design, and construction; sample collection, handling, and laboratory analysis; analysis of detection-monitoring data; and elements of an effective long-term monitoring program. The *TEGD* is not intended to serve as a set of regulations; rather, it is provided to evaluate the adequacy of the monitoring programs for contaminants associated with RCRA and non-RCRA facilities.

The EPA's *Practical Guide for Ground-Water Sampling* complements the content of the *TEGD* because effective groundwater monitoring procedures are described in significantly more detail. This publication is of value to field personnel and environmental professionals during the development and execution of a groundwater monitoring program. Although somewhat similar to the *TEGD*, the *Practical Guide for Ground-Water Sampling* (Barcelona et al., 1985) emphasizes quality assurance (QA), quality control (QC), field techniques, specific sampling protocol, and data evaluation.

The ASTM has assembled a series of standard practices pertinent to groundwater monitoring activities. ASTM Designation D4448-85a, *Standard Guide for*

Sampling Groundwater Monitoring Wells, is intended to serve as a guide to the most commonly used methods for properly collecting, handling, and preserving a groundwater sample for laboratory analysis. ASTM Designation 5092-90, *Standard Practice for Design and Installation of Groundwater Monitoring Wells in Aquifers,* provides standard protocol to ensure that the water extracted from the monitoring well is representative of the targeted geologic formation. Various site-characterization strategies, drilling methods, monitoring well materials, and installation techniques are presented in an effort to minimize sampling errors during the monitoring program. ASTM PS64-96, *Provisional Standard Guide for Developing Appropriate Statistical Approaches for Groundwater Detection Monitoring Programs,* provides statistical methods to filter out false positive or false negative analytical results. ASTM has also prepared numerous other documents that are directly or indirectly related to groundwater monitoring and soil-sampling techniques that will be referred to during this discussion.

HYDROGEOLOGIC PROCESSES

Groundwater contamination is generally defined as the presence of a chemical or solute in the groundwater at such a concentration that the resource is no longer suitable for a specific use. Note that groundwater contamination may be the result of human activity or a natural occurrence. Examples of naturally occurring groundwater contamination include a high total dissolved solids content or dissolved metals in the groundwater originating from minerals present in the matrix of the aquifer. Unfortunately, examples of groundwater contamination resulting from human activities are plentiful. Most commonly, underground storage tanks (USTs), septic tanks, uncontrolled landfills, surface spills of chemicals, mine tailings, injection wells, waste treatment or disposal lagoons, application of agricultural chemicals, and improperly constructed water, oil, and gas wells are cited as sources of groundwater and soil contamination.

It is necessary to employ a long-term groundwater monitoring program for contaminated sites because groundwater conditions change in response to physical, biological, and geochemical forces acting on the groundwater contained within an aquifer. Thus, the concentration of a solute changes with time as a function of the processes that influence the movement and chemistry of the groundwater. A groundwater monitoring event therefore depicts the conditions that exist only at a particular point in time and is analogous to a snapshot. To characterize the processes that influence the contaminant plume, the results of several monitoring events must be analyzed to understand how contaminant concentrations change with time.

The vertical infiltration of a contaminant through the surface soil may occur as pure product or as a dissolved constituent as precipitation leaches the more soluble compounds from the contaminated surface soils. Once the contaminant percolates, or infiltrates, to the groundwater, chemical products that are less dense than water (i.e., most petroleum-based fuels) tend to float on the water table. Chemicals that are denser than water (some chlorinated solvents) continue to migrate vertically through the water-bearing horizon and accumulate on top of less permeable units within or at the base of an aquifer. In either case, the solubility of a chemical species dictates how much solute is available for transport by groundwater movement from the source area.

The primary physical processes that influence contaminant concentration within an aquifer are advection, diffusion, and dispersion. *Advection* is usually the most significant factor in the movement of contaminated groundwater through an aquifer because the dissolved chemical is transported from the source area as the groundwater flows through the subsurface to areas of lower hydraulic head. The velocity of groundwater movement through an aquifer is dependent on the porosity, hydraulic conductivity, and gradient of an aquifer. It can commonly range from a fraction of a centimeter/day to several meters/day, depending on site-specific conditions. *Diffusion* is the process through which dissolved chemical species migrate from areas of high concentration to areas of lower concentration. Contaminant velocities due to diffusion are generally orders of magnitude less than those of advection in an aquifer. Together, advection and diffusion are the transport mechanisms that directly affect contaminant distribution.

Dispersion is a function of the porosity of the aquifer and the flow path of the water around the aquifer matrix. Each of these physical processes acts to transport the solute from the source area in a direction generally parallel to the direction of groundwater flow, resulting in the formation of a *groundwater contaminant plume*, as shown in Figure 9–1.

The biological processes that act to reduce the subsurface contaminant concentration through time are dependent on nutrient supply, microorganism species, and the toxicity of the contaminant released. The soil profile is normally occupied by an abundance of microorganisms that metabolize naturally occurring minerals and organic matter to further enrich the soil. Microbes are also readily able to metabolize many hydrocarbon compounds, as long as the contaminant concentrations are not toxic and the other nutrients necessary to sustain growth are not limiting the microbial population. Therefore, significant microbial degradation of hydrocarbons is commonly observed during long-term monitoring programs.

The principal chemical processes that affect contaminant concentrations are degradation and retardation. *Degradation* of a contaminant may occur as the chemical reacts with the aquifer matrix or the chemical characteristics of the groundwater itself. *Retardation* is a function of the interaction between the aquifer matrix and the dissolved chemical. Ion exchange, precipitation, and sorption due to clay mineralogy and organic carbon content are important components of chemical retardation.

The combined effect of these processes produces a groundwater contaminant plume that appears to migrate at a rate somewhat slower than the calculated average linear velocity of the groundwater within the aquifer. Chemical species are transported through the subsurface at different rates, depending on the characteristics of that chemical. A series of monitoring well measurements and analytical results typically shows the arrival of the more soluble and mobile chemical compounds, followed by progressively less soluble species. Therefore, natural processes act collectively to segregate the mixture of chemicals in a fuel spill from the moment it is released into the environment. As the most volatile fraction escapes to the atmosphere before infiltration into the subsurface, a portion of the spill is sorbed to soil particles and organic carbon, microbial degradation metabolizes a portion of the release, and groundwater advection transports the dissolved constituents away from the point of the release at differing velocities, depending on the retardation characteristics of each chemical and the aquifer matrix.

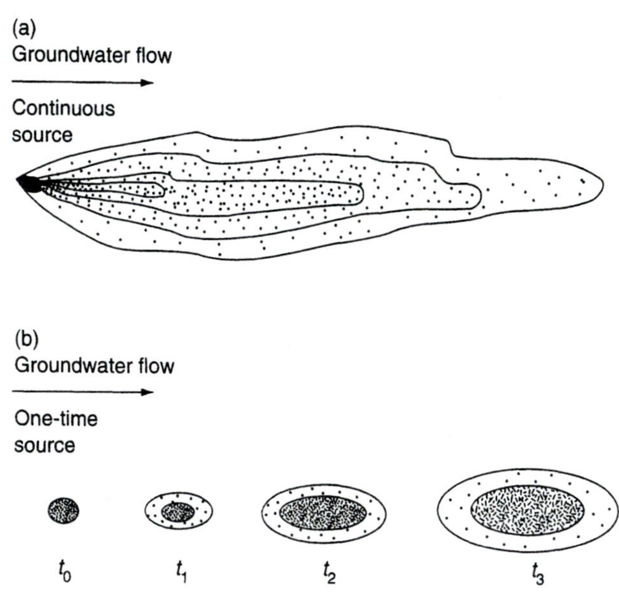

FIGURE 9–1
The development of two theoretical groundwater plumes as a result of (a) the release of the contaminant to the environment from a continuous source, such as gasoline released from a leaking UST, and (b) the intermittent release of the contaminant to the subsurface, as might occur with illegal dumping activities (from Fetter, 1988).

GROUNDWATER MONITORING PROGRAMS AND OBJECTIVES

Groundwater monitoring programs are implemented to provide data for various objectives. As Fetter (1988) noted, the following are the most common monitoring program objectives:

1. Evaluate the groundwater quality from a specific aquifer or water supply well.
2. Evaluate the potential for hazardous substances that have negatively impacted property value.
3. Delineate the extent or severity of contamination from an identified contaminant source area.
4. Detect the migration of contaminants toward a sensitive receptor.

Four levels of groundwater monitoring programs may be employed, depending on the intent of a particular investigation. Each program is developed based on the available site data and the judgment of the environmental professional supervising the program. Because the levels of monitoring basically represent a progression of monitoring activities through the evolution of a site, from discovery through formal closure, the information available at each stage can be used to further refine the program and focus on indicator parameters or specific chemical species of concern.

The primary types of groundwater monitoring programs can generally be classified into the following:

1. Detection sampling.
2. Assessment sampling.
3. Long-term monitoring.
4. Closure sampling.

Detection sampling programs are developed with limited site-specific groundwater quality data because the results of this investigation usually serve as the baseline for further analytical work. Detection sampling is employed in environmental site assessments to evaluate the potential for soil or groundwater contamination prior to property transfer. A detection monitoring program also is appropriate for the evaluation of the water quality for a specific well, aquifer, sensitive receptor, or area adjacent to a waste storage/disposal facility.

A broad scope of analytical data is generally involved with the execution of a detection sampling program. The critical elements that must be considered during the development of a detection monitoring program are identifying environmental concerns, appropriate monitoring well placement and design, acquiring representative samples, and obtaining statistically meaningful laboratory results. Each of these criteria is discussed in greater detail through the balance of this chapter.

The need for an *assessment sampling* program typically results from the discovery during the detection sampling phase of contaminant concentrations that exceed state or federal corrective action levels. Generally, the state or federal regulatory agency dictates which parameters warrant further sampling to properly delineate the extent of contamination. Additional information that may be obtained during the assessment phase includes the rate of contaminant migration, delineation of horizontal and vertical extent of contamination, contaminant concentration gradients within the impacted area, and identification and sampling of potential receptors.

Long-term monitoring programs are implemented after the assessment phase is complete and the site has been characterized to the satisfaction of the governing agency. Such a program may be implemented during corrective action if the contamination warrants remediation to monitor for off-site migration and the general progress of the remedial effort. If the contaminant concentrations do not justify remedial action, a long-term monitoring program may be an appropriate response to document the rate of natural attenuation. Because the assessment phase has served to document the contaminants of concern, the scope of work for the long-term monitoring analytical program can focus on specific chemicals to reduce overall costs.

When the long-term monitoring results indicate that the objectives of the corrective action plan have been attained, the owner of the site may proceed with *closure sampling* to document that the levels requiring corrective action no longer exist and site remediation is complete. The Closure Plan may require that additional monitoring wells be installed within the existing network of monitoring points to ensure that the remediation is complete. Regulators commonly request additional monitoring episodes to detect the reoccurrence of contaminants as the aquifer returns to static conditions upon termination of the remediation effort. Therefore, closure sampling has two main objectives: first, to document that the corrective action has been effective and, second, to detect remobilization of contaminants as the aquifer returns to its natural state.

Site Assessment Monitoring Goals

The primary goal of the site assessment (or site characterization) monitoring phase is to collect sufficient data to adequately describe the hydrogeologic regimes that may potentially be impacted by the site-specific contaminants. To adequately characterize a site, the environmental professional must describe the regional groundwater flow characteristics, be certain to identify the uppermost water-bearing horizon and deeper aquifer units that may be impacted, delineate the horizontal and vertical extent of the contaminant plume, understand how seasonal variability may affect the data obtained, and detect pathways of preferential groundwater flow. A network of assessment monitoring wells is installed to demonstrate adequate understanding; selected soil and groundwater samples are then submitted to the laboratory for analysis of targeted contaminants.

A minimum of three to four groundwater monitoring wells are necessary to characterize a single water-bearing unit at a suspected release site. However, sites that are adequately characterized by three or four monitoring points are quite rare, because most environmental investigations require many more monitoring points (commonly six to twelve wells for a relatively simple site) to adequately evaluate the horizontal and vertical extent of soil and/or groundwater contamination. Sites that exhibit complex geology (fractures, lithologic discontinuities, facies changes, tilted or folded beds, zones of enhanced permeability, etc.) usually require a greater number of monitoring wells to be adequately characterized (*TEGD*, 1986).

The assessment drilling strategy is extremely important to the success of the subsequent monitoring program because the network of monitoring wells installed during this phase of work must serve as the baseline data set and provide an accurate representation of the aquifer conditions during future sampling events. There are at least five fundamental strategies that must be considered for the proper placement of assessment monitoring points (Lipson and Roe, 1991):

1. Define the horizontal and vertical areas of the investigation based on the characteristics of the suspected contaminants and the site geology.
2. Estimate the direction of groundwater movement. Groundwater, like surface water, tends to follow the path of least resistance. The topographically downslope (down-gradient) direction can generally be used at this stage of the investigation as a guide to estimate the direction of groundwater flow. Be aware that stratigraphic controls and utility conduits may significantly influence the local groundwater flow direction.
3. The investigation should adequately evaluate possible areas of petroleum or hazardous substance storage, use, or disposal. Visual observation, personnel interviews, site records, fire insurance maps, and Phase I Environmental Site Assessments may be used to identify areas of concern.
4. Establish at least two monitoring wells in the anticipated down-gradient direction from each area of concern. One of these wells should be located as close to the source area as possible. The other well may be positioned 20 to 50 feet in the down-gradient direction, depending on the sediment type observed in the first boring.
5. Establish a monitoring well in an up-gradient (topographically up-slope) direction from each area of concern. The data from this monitoring point provide valuable background water quality information, assist in identifying or discounting

up-gradient sources possibly impacting the area of concern, and provide necessary groundwater elevation information to determine the direction and velocity of groundwater movement.

WORK PLANS

Work plans are necessary to successfully execute an effective groundwater monitoring program. Depending on the complexity of the site and the requirements of the property owner and the regulatory agency, the content of the required plans may be adequately presented within several succinct paragraphs or may comprise volumes of detailed procedures and protocol.

Quality Assurance and Quality Control Plan

A quality assurance program is a documented system of checks and procedures used to verify that the field and laboratory data obtained are reliable. Quality control procedures are intended to reduce variability, for example, through the analysis of duplicate and blank samples.

The goal of every *Quality Assurance and Quality Control (QA/QC) Plan* is to ensure that the hydrogeologic and analytical data collected are accurate, precise, complete, and reproducible. The QA/QC Plan defines the minimum level of analytical sensitivity that will meet the objectives of the investigation. *Accuracy* is generally defined as the degree of deviation from a true value and can be evaluated through the analysis of spiked samples by which a known amount of contaminant is introduced into a sample. *Precision* is the probability that a particular analysis of duplicate samples will reside within predefined confidence limits from a mean value. The standard procedures outlined in the QA/QC Plan will improve the reproducibility of the monitoring data because strict adherence to standard techniques will reduce the chance for random, systematic, and analytical errors. Sensitivity is a function of the analytical method detection limit to attain reliable measurements and the smallest concentration of chemical that is detectable with a particular laboratory instrument (Barcelona et al., 1985).

Groundwater Sampling and Analysis Plan

The *Sampling and Analysis Plan* is a critical component of any groundwater sampling program. This document serves as the written protocol to ensure that the appropriate procedures are followed consistently during each groundwater sampling episode. The Sampling and Analysis Plan provides detailed instructions for sample collection, preservation, and shipment. In addition, it identifies the specific laboratory procedures to monitor the presence of targeted contaminants. Chain-of-custody procedures are also presented in the Sampling and Analysis Plan in an effort to prevent or detect tampering. Detailed QA/QC procedures for field instrumentation also may appear.

Health and Safety Plan

The hazard communication requirements of the Occupational Safety and Health Administration (OSHA) dictate that a site-specific *Health and Safety Plan* be prepared for sites subject to intrusive sampling programs where there is a potential for worker exposure to chemical hazards. The Health and Safety Plan must identify the possible contaminants present, exposure symptoms, emergency procedures, site control, decontamination procedures, and personal protective equipment, in addition to other issues of concern.

DRILLING TECHNIQUES AND ASSESSMENT STRATEGIES

A variety of methods can be employed during an environmental investigation to install a monitoring well. There are certain advantages and disadvantages to each technique. This section briefly summarizes the most commonly employed drilling techniques for environmental investigations. The following criteria should be considered when selecting the most appropriate drilling method:

- Quality of soil samples obtained during drilling.
- Ease of well installation and anticipated water production.
- Desired well diameter and depth of casing.
- Access to the site, slope, and surface conditions.
- Provisions to guard against cross-contamination.
- Site geology, potential to encounter bedrock.
- Cost, size, and efficiency of drilling crew.

Continuous-Flight Auger Systems

The hollow-stem *continuous-flight auger system* is the most common type of drilling tool used for environmental investigations in unconsolidated sediments. The primary advantages of this system include the ability to collect relatively undisturbed soil samples and the minimal risk of cross-contamination because the hollow-stem augers provide for the retrieval of discrete samples with a core barrel that is raised and lowered through the center of the augers. Hollow-stem augers provide a temporary casing to insert the monitoring well casing once the boring has reached the targeted horizon. During well construction, the hollow-stem augers are extracted one section at a time to prevent the sides of the boring from collapsing into the open borehole as the filter pack, bentonite, and grout are positioned around the annulus of the casing. The typical configuration of a hollow-stem auger system is depicted in Figure 9–2.

The solid-stem continuous-flight auger system is not commonly used in environmental investigations. It is appropriate only where the geologist already has a good understanding of the sediments present and the contaminant distribution at the site. If the sidewalls of the boring do not collapse when the augers are completely withdrawn, a monitoring well may then be constructed in the open borehole. This drilling technique is quicker and less costly than the hollow-stem method; however, the soil samples are highly disturbed and homogenized by the time the cuttings are carried to the surface by the auger flights.

The depth limitation of continuous-flight auger systems is generally around 40 to 50 meters. These techniques are appropriate only in unconsolidated sediments. Large cobbles and boulders will usually present significant drilling problems.

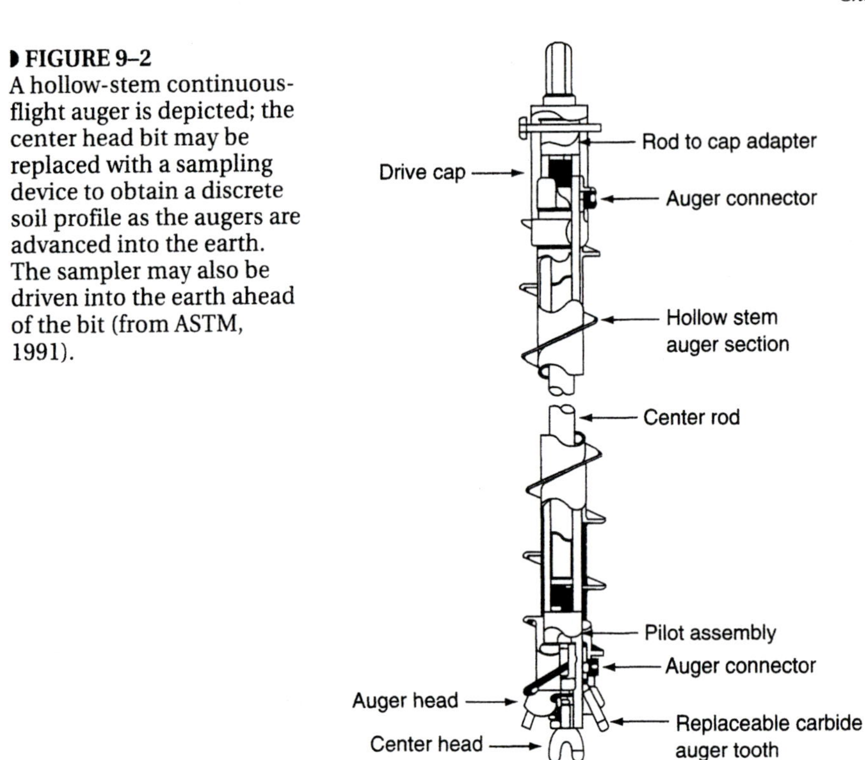

FIGURE 9–2
A hollow-stem continuous-flight auger is depicted; the center head bit may be replaced with a sampling device to obtain a discrete soil profile as the augers are advanced into the earth. The sampler may also be driven into the earth ahead of the bit (from ASTM, 1991).

Rotary Drilling Techniques

For deeper borings and consolidated formations, it may be necessary to use *rotary drilling* techniques to construct a monitoring well. With rotary drilling, a circulating fluid (water or drilling mud) or compressed air is pumped down the center of the drill pipe, exits at the bottom of the boring, flows around the drill bit at a high velocity, and then is displaced back to the surface within the annulus between the exterior of the drill pipe and the sidewalls of the boring. The circulating fluid serves an important purpose because it transports cuttings from the boring, cools the bit, and helps to support the sidewalls of the boring.

Rotary drilling techniques are not commonly used in environmental investigations because of their reduced sample quality and greater expense. The geologic conditions that might make this a valid alternative are not usually encountered because most environmental projects investigate only unconsolidated sediments.

Bucket Auger Drilling Rig

The *bucket auger drilling rig* uses a large diameter cylindrical bucket up to 1.5 meters tall and equipped with a series of chisel teeth on a hinged trap door to drill large diameter wells up to 2 meters in diameter. As the drill string (kelly rod) is turned, soil cuttings accumulate in the bucket. When the bucket is full of cuttings, it is extracted from the boring and pulled to the side, releasing the trap door and allowing the cuttings to be emptied on the surface. A typical bucket rig is shown in Figure 9–3.

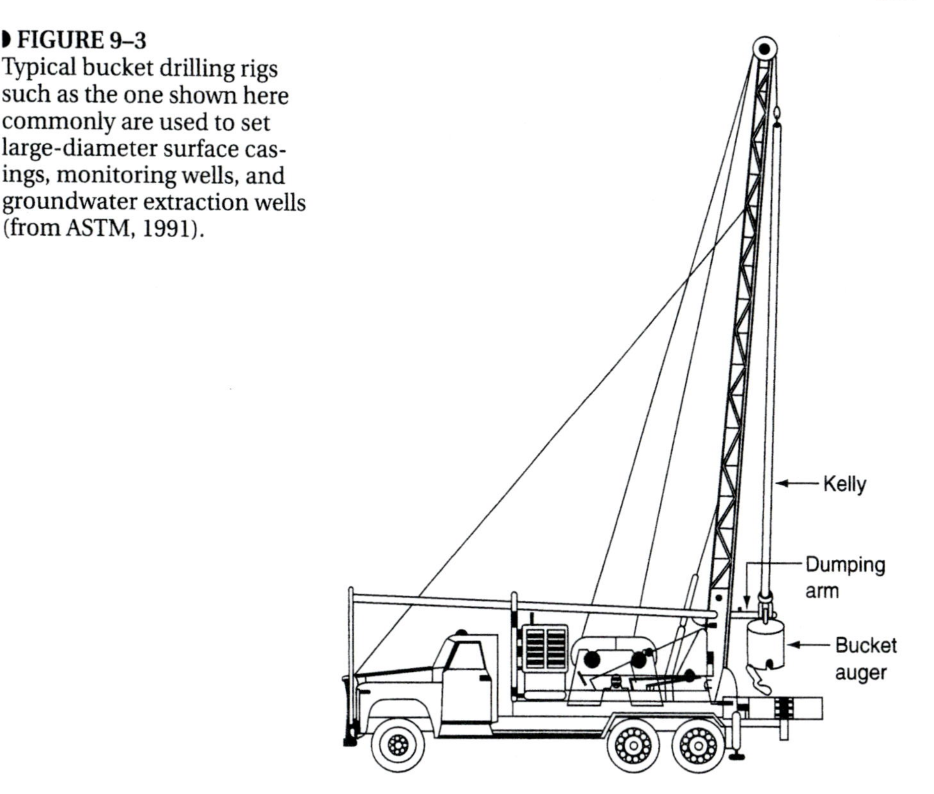

FIGURE 9–3
Typical bucket drilling rigs such as the one shown here commonly are used to set large-diameter surface casings, monitoring wells, and groundwater extraction wells (from ASTM, 1991).

The environmental applications for this method generally apply to installation of shallow groundwater extraction wells in fine-grained unconsolidated sediments (silt and clay). Fairly undisturbed soil samples may be collected from the cuttings after the bucket has been dumped.

Direct-Push Drilling Techniques

A variety of direct-push drilling techniques exists; they use hydraulic pistons or percussion hammers to drive well casing, and screened sections to collect samples from a targeted depth. Recently, some techniques have been devised to retrieve a discrete core of sediment for logging and sampling purposes.

The primary advantages of this technique are reduced costs, improved site access, and speed. In addition, discrete samples may be collected in acetate liners to prevent cross-contamination. Direct-push techniques work well in fine-grained sediments and shallow water table conditions.

BORING LOG CONTENT

The most important source of geologic data for an environmental investigation is the boring log that is prepared in the field during the drilling and well installation process. The field geologist records the soil/rock characteristics, depth, instrument readings, and other significant details in a field notebook or on a *boring log*. The boring log typically presents the following information at a minimum (ASTM, 1993):

1. **Project information.** Site name, location, elevation, drill crew, and geologist's name and company.
2. **Exploration information.** Boring number, drilling method, drill make and model, weather conditions, date started/finished, total depth and diameter of boring, casing details, drilling conditions, and sampling method.
3. **Subsurface information.** Depth and petrographic description of soil/rock units encountered using a standard classification scheme, field test results (organic vapor monitor, pH, pocket penetrometer readings, etc.), and sample intervals for laboratory analysis.
4. **Groundwater information.** Method of measurement, depth to groundwater, and field test results.
5. **Monitoring well construction information.** Casing material, diameter and length, screen type and opening size, screened interval, type of filter pack, bentonite plug type and position, backfill material, grout type and interval, protective cover and well cap, and elevation of top of casing.

The standard techniques for describing soils are presented in ASTM Standard D2488-93, *Standard Practice for Description and Identification of Soils* (1993). The soil classification system presented in ASTM D2488-93 is based on visual examination and manual tests that can be conducted in the field to describe the physical characteristics of the soil sample.

The data presented in the boring logs provide the foundation for geologic interpretations and developing an understanding of subsurface processes affecting contaminant distribution. The boring logs should always be carefully reviewed because they provide the basis for preparing geologic cross sections, estimating the volume of contaminated soil and groundwater, performing hydraulic conductivity calculations and the identification of sensitive aquifer units, and confining beds that may restrict contaminant movement.

MONITORING WELL DESIGN AND INSTALLATION

The primary purpose of the monitoring well is to facilitate the collection of water samples and field data representative of the surrounding formation. Proper monitoring well design and installation techniques are integral components of an effective groundwater monitoring program. To properly design a monitoring well, the geologist must first develop a conceptual model of the hydrogeologic horizon that is to be sampled and understand the interrelationship between that unit and surrounding formations. An understanding of the chemical characteristics of the contaminant of concern, seasonal water table elevations, and effect of nearby water wells, rivers, paved areas, and such, are considered during the development of the conceptual hydrogeologic model.

ASTM Designation D5092-90, *Standard Practice for Design and Installation of Ground Water Monitoring Wells in Aquifers*, provides detailed guidance on the proper drilling methods, well construction materials, sizing, and installation techniques for typical groundwater monitoring wells. The essential components of a monitoring well consist of the sand pack, well screen, riser, annular sealants, and a surface protector (see Figure 9–4). (The following discussion provides only a general description of the components of a typical monitoring well. For additional information, refer to ASTM D5092-90 or the *TEGD*).

The sand pack (see Figure 9–4) consists of properly graded sand or gravel that is placed in the annulus around the well screen to support the sidewall of the surrounding formation and prohibit fine-grained sediment, carried by groundwater, from entering the well. The sand pack gradation should be fine enough to capture fine-grained sediment from the formation, but coarse enough so that it will not be washed into the monitoring well casing. The sand pack should extend from the base of the screened interval to no more than 2 feet above the top of the screen. ASTM D5092-90 provides guidance on selecting the proper gradation of sand pack based on site-specific conditions.

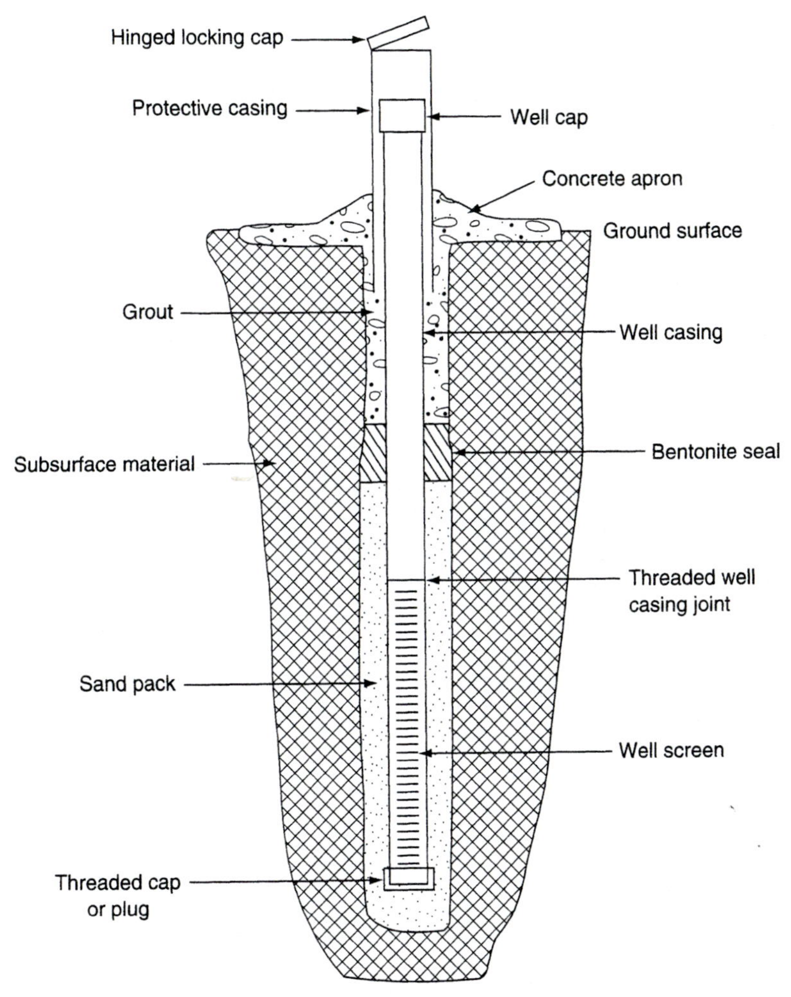

FIGURE 9–4
Typical monitoring well construction diagram (from Gates, 1989).

A variety of well screen is commercially available. Most commonly, the well screen is composed of polyvinyl chloride (PVC), fiberglass, or stainless steel. The type of material selected should be based on compatibility with the contaminant of concern, cost, and durability. The length and position of the screened interval should be determined based on site-specific conditions. If the contaminant is lighter than water (a light nonaqueous phase liquid [LNAPL]), the screened interval should generally straddle the water table and allow enough screen above the water level to accommodate seasonal fluctuations of the water table. If the contaminants are denser than water (a dense nonaqueous phase liquid [DNAPL]), the screened section should be located at the base of the aquifer because dense liquids tend to accumulate on the surface of the underlying *aquitard* (a confining unit that retards the movement of groundwater). Refer to Figure 9–5, which depicts the proper well-screen locations to detect nonaqueous phase liquids of various densities.

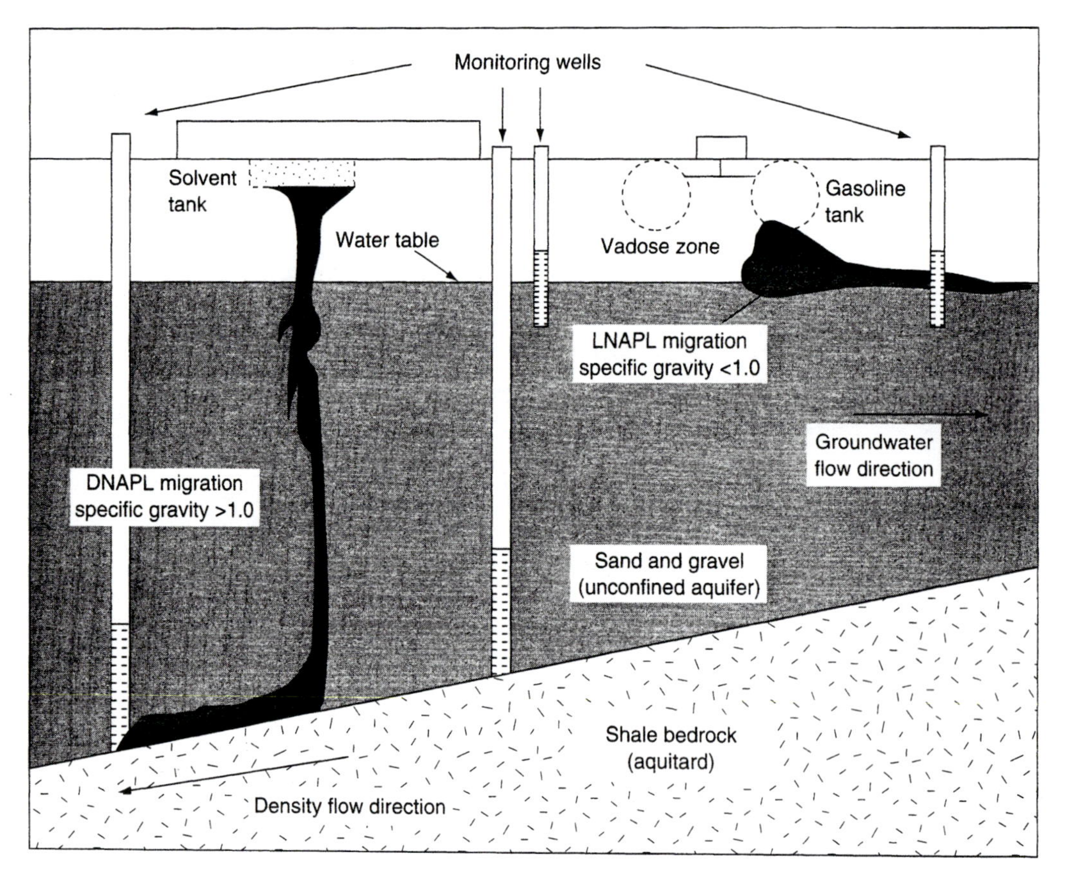

Figure 9–5
Diagram showing the placement of groundwater monitoring wells in relation to a LNAPL and DNAPL release. Note that the LNAPL migrates down-gradient with respect to the water table elevation while the DNAPL moves in the downslope direction on top of an aquitard (after McLelland and DeSocio, 1991).

The proper slot size opening for the screen also must be selected based on site-specific characteristics. The width of the slot must retain at least 90 percent of the filter pack placed around the annulus of the casing to prevent the filter pack and formation material from being washed into the well casing.

The riser is usually constructed of the same material as the well screen. As with all monitoring well construction materials, the riser should be free of grease, cutting oil, plastic solvents, and glues. The riser must be of sufficient strength to withstand the tensile stress of the installation process and the compressive force of the surrounding formation.

The purpose of the annular sealants is primarily to prevent the vertical migration of water along the annulus surrounding the well casing. Surface water (and contaminants) may migrate into the monitoring well if the annular seal does not provide an adequate seal. Bentonite, a montmorillonite clay with high swelling capacity upon hydration, is commonly used as an annular sealant to isolate the screened interval from cross-contamination or to plug abandoned borings and monitoring well casings. The bentonite seal is usually 1 to 2 meters thick, at a minimum, and placed directly on top of the filter pack. Grout, a bentonite-Portland cement-water mixture, is often placed on top of the bentonite annular seal to just below the frost line around deep well casings in arid regions. Concrete is usually placed from the surface to below the frost line to prevent the well from heaving.

The typical monitoring well is then equipped with a surface protector, watertight well cap, and lock to limit access. Surface protectors can be mounted flush with the surface in areas of traffic and set in concrete. The preferred technique, however, is to extend the well casing and protector approximately 1 meter above grade. This construction method ensures that surface water and surface spills do not directly enter the well casing.

GROUNDWATER SAMPLE COLLECTION PROTOCOL

Standard groundwater sample collection procedures include the following sequence of events:

- Monitoring well development.
- Measuring water level.
- Purging wells.
- Acquiring samples.
- Handling and preserving samples.
- Completing a chain-of-custody form.
- Delivering samples to the analytical laboratory.

This section provides a conceptual discussion of each step. Worker health and safety are of paramount importance. Before accessing any environmental monitoring well, the field technician must review the site-specific health and safety plan to understand potential exposure symptoms, personal protective equipment, evacuation procedures, decontamination methods, and site hazards.

Note: For a detailed description of sampling techniques, refer to the site-specific sampling and analysis plan, QA/QC plan, and to ASTM 4448-85(a)—*Standard Guide for Sampling Groundwater Monitoring Wells.*

Monitoring Well Development

The purpose of well development activities is to remove drilling fluids and sediment introduced into the well during the drilling process. Another objective is to restore the ability of the aquifer to transmit groundwater that is representative of the formation to the well casing. The hydraulics of the well are improved considerably after the development process as fine-grained sediment is removed from the well and drill cuttings smeared on the sidewalls of the formation are loosened.

The basic technique of well development involves removing groundwater from the well casing at an increasingly more vigorous rate. As formation water flows into the evacuated casing, it loosens fine-grained sediment and cuttings within the filter pack, which are washed into the casing and subsequently removed. On completion of well development activities the formation should provide relatively sediment-free (nonturbid) water. The yield of the aquifer also will be enhanced by proper well development, facilitating a representative estimation of hydraulic conductivity.

A variety of techniques can be used to develop a groundwater monitoring well. A *surge block* acts similar to a piston driving water back and forth across the filter pack to loosen sediment and drill cuttings; a bailer is then used to withdraw the sediment-laden water from the casing. Vigorous pumping in conjunction with surging will rapidly develop a monitoring well. The gentle action of an air-lift pump will effectively remove fine-grained sediment from the filter pack. The well may also be subjected to *jetting*, with a pressurized water source of known chemistry used to free material lodged in the well screen. These techniques are presented in more detail in ASTM D5092.

Water Level Measurements

To prepare meaningful hydrogeologic maps and cross sections, the horizontal and vertical positions of each boring and monitoring well must be accurately surveyed. The elevation of a reference point (notch cut into the casing) on the top of the well casing must be tied to a known benchmark. Depth to water measurements are always measured from the reference point cut in the well casing to the static water level in the well. The elevation of the water table (*potentiometric surface*) in the well can then be calculated. The potentiometric surface is the elevation at which the water pressure head in the aquifer is equal to atmospheric pressure.

The elevation of the potentiometric surface at each monitoring point should be plotted on a scaled base map of the site. Standard contouring techniques can then be used to contour *equipotential lines* (groundwater elevation contour lines) to determine the direction of groundwater flow through the aquifer. A minimum of three potentiometric surface elevations is required to determine groundwater flow direction. The difference in groundwater elevation as a function of horizontal distance provides a measure of the hydraulic gradient at the site, which is one of the parameters needed to estimate the velocity of groundwater through the aquifer.

A variety of devices may be used to obtain depth-to-water measurements, depending on the specific objectives of the Sampling and Analysis Plan. A steel tape, coated with a special paste, commonly has been used to measure the distance from the reference point cut into the riser to the water level in the well. Electrical conductivity, light reflectance, and acoustical-based electronic water level indicators also are widely used devices that present certain advantages over the steel tape method. Some conductivity-based water level indicators also are capable of detecting non-aqueous phase liquids (NAPL) floating on the water table or accumulations at the base of the aquifer, depending on the specific gravity of the liquid. If LNAPL is noted in a well, the water table elevation should be corrected to compensate for the isostatic effect of the floating product.

Well Purging

Because the groundwater in the well casing may not be representative of the formation water after sitting for a long period of time, it is important to require specific well-purging techniques in the Sampling and Analysis Plan. Well purging is intended to gently evacuate the groundwater from the well casing and the surrounding sand pack and to allow the groundwater to be drawn into the well to obtain a representative sample. It is standard practice to withdraw three to five saturated casing volumes from the well and allow the well to recover its static elevation prior to collecting a sample. Physical characteristics (temperature, pH, and/or conductivity) of the groundwater extracted from the well also may be monitored to indicate when the well is adequately purged. Consistent well-purging techniques are extremely important during long-term monitoring programs to obtain comparable and reproducible data throughout the duration of the monitoring period.

A variety of devices may be used to purge and sample a groundwater monitoring well. The most common and economical device is the *disposable bailer*, which may be discarded after sample acquisition or dedicated to the monitoring well for future purging/sampling events. A bailer is simply a tube of inert material (e.g. PVC, stainless steel, or acrylic) equipped with a check valve in one (or both ends) of the device. A string composed of inert material (such as cotton) may be used to repeatedly lower the bailer into the well to evacuate the necessary volume of groundwater. Other purging devices commonly employed include submersible electric pumps, pneumatic bladder pumps, suction lift pumps, and piston pumps. The appropriate sampling device is selected based on the chemical parameters being monitored, site conditions, and cost (ASTM, 1992). The purge water accumulated at the surface is discharged into a 5-gallon bucket and disposed of according to the protocol described in the Sampling and Analysis Plan.

The labor and disposal costs involved in purging the required volume of groundwater from deep wells can become significant burdens. For this reason, some recent Sampling and Analysis Plans have specified low-flow well-purging techniques. This method involves extracting groundwater with an adjustable low-flow pump from near the middle of the saturated interval of the well screen at a rate below the recharge capacity of the formation. Pumping at a rate of less than 1 liter/minute is usually sufficient to obtain a representative sample after certain indicator parameters (pH, temperature, conductance, reduction-oxidation potential, dissolved oxygen, and turbidity) have stabilized (Puls and Barcelona, 1995).

LABORATORY ANALYTICAL METHODS

Soil and groundwater samples collected during an environmental investigation should be analyzed for the suspected constituent suite according to approved state or EPA laboratory methods to report valid and meaningful results. Specific laboratory methods must be identified in the Sampling and Analysis Plan to provide the desired sensitivity for the property owner and/or regulatory agency to make recommendations and determine an appropriate course of action.

Physical characteristics of soil/rock samples should be determined using appropriate ASTM methodologies. Such physical characteristics as grain size distribution, porosity, and hydraulic conductivity provide valuable data to determine the hydrogeologic behavior of the geologic units that influence contaminant migration.

Chemical characteristics of the soils that are important in an environmental assessment of an impacted site include organic carbon content, mineralogy, and contaminant concentration. The data obtained from these analyses will provide information to calculate the retardation coefficients and contaminant migration velocities.

EPA document SW-846, *Test Methods for Evaluating Solid Waste, Physical/Chemical Methods,* describes, in detail, the analytical procedures to evaluate groundwater samples for the presence of the pollutants regulated under RCRA. Federal and state action levels that are to be protective of drinking water resources have been established for these contaminants. Generally, the maximum contaminant level (MCL) for drinking water as defined by the Safe Drinking Water Act serves as the enforceable action level for groundwater contaminants. Therefore, if the MCL is exceeded for a groundwater sample, corrective action will be required by the regulatory agency. Table 9–1 shows a partial listing of the MCLs for commonly encountered volatile organic compounds.

TABLE 9–1
Maximum contaminant levels of selected volatile organic compounds.

Chemical Name	MCL (mg/l)
Benzene	0.005
Carbon tetrachloride	0.005
Chlorobenzene	0.100
o-Dichlorobenzene	0.600
p-Dichlorobenzene	0.075
1,2-Dichloroethane	0.005
1,1-Dichloroethylene	0.007
cis-1,2-Dichloroethylene	0.070
1,2-Dichloropropane	0.005
Ethylbenzene	0.700
Styrene	0.100
Tetrachloroethylene	0.005
Toluene	1.000
1,1,1-Trichloroethane	0.200
Trichloroethylene	0.005
Vinyl chloride	0.002
Xylenes, total	10.000

Source: EPA Drinking Water Standards, January 1996.

HYDROGEOLOGIC CHARACTERIZATION

The collection of groundwater monitoring data facilitates the development of contaminant distribution maps, groundwater gradient maps, estimations of hydraulic conductivity, and calculations of solute velocity in the subsurface. Collectively, these data and interpretations constitute the groundwater characterization for a site.

Hydraulic conductivity testing is performed after the well has been properly developed to determine the rate at which groundwater will move through the aquifer. These data, coupled with the effective porosity of the formation, hydraulic gradient, and retardation factors, can be used to estimate the rate of contaminant migration through the aquifer. To determine hydraulic conductivity, the hydrogeologist must be careful to use the appropriate analytical solutions that are valid with respect to the site-specific conditions.

Slug tests can be used to rapidly measure the response of the aquifer immediately surrounding the monitoring well to the sudden introduction, or withdrawal, of a slug of known volume. The rate at which the water level in the well then equilibrates provides the basis for calculating the average hydraulic conductivity across the screened interval. *Pumping tests* are conducted to obtain a more accurate evaluation of the hydraulic conductivity of the area (within several meters) around the extraction well. During several hours of pumping water at a constant rate from an extraction well, water level measurements are collected from nearby observation wells to measure the response of the aquifer to the groundwater withdrawal. Hydraulic conductivity can then be determined using a graphic technique to match the observation well data plot and standard response curves.

Porosity (the ratio of void spaces in the sediment to the total volume of the sample) of the aquifer matrix can be determined in the laboratory or by using reference tables based on lithology such as Figure 9–6. An estimation of the *specific yield* (porosity of a saturated sample that will yield water due to gravity drainage) is needed to calculate the average linear velocity of groundwater through the aquifer.

STATISTICAL ANALYSIS OF MONITORING DATA

To better define sources of monitoring variability and derive an optimum sampling frequency for a long-term monitoring program, Barcelona et al. (1989) conducted a monitoring program with stringent QA/QC procedures on a site for a period of two years. Their results indicated that there are generally three sources of variability in a monitoring program that can have statistically significant effects. The first, natural variability, can be attributed to geochemical processes, water table fluctuations, and seasonal temperature ranges. The second, monitoring network design and operation, may create another component of variability, resulting from improper well-purging techniques, sampling methodology, decontamination procedures, and other field activities. The third source of variability may be due to analytical errors and is easily monitored through an enhanced QA/QC Plan. Cumulatively, these sources of variability resulted in a deviation of approximately 20 percent from mean indicator parameter values. These results underscore the importance of consistent and thorough well-purging

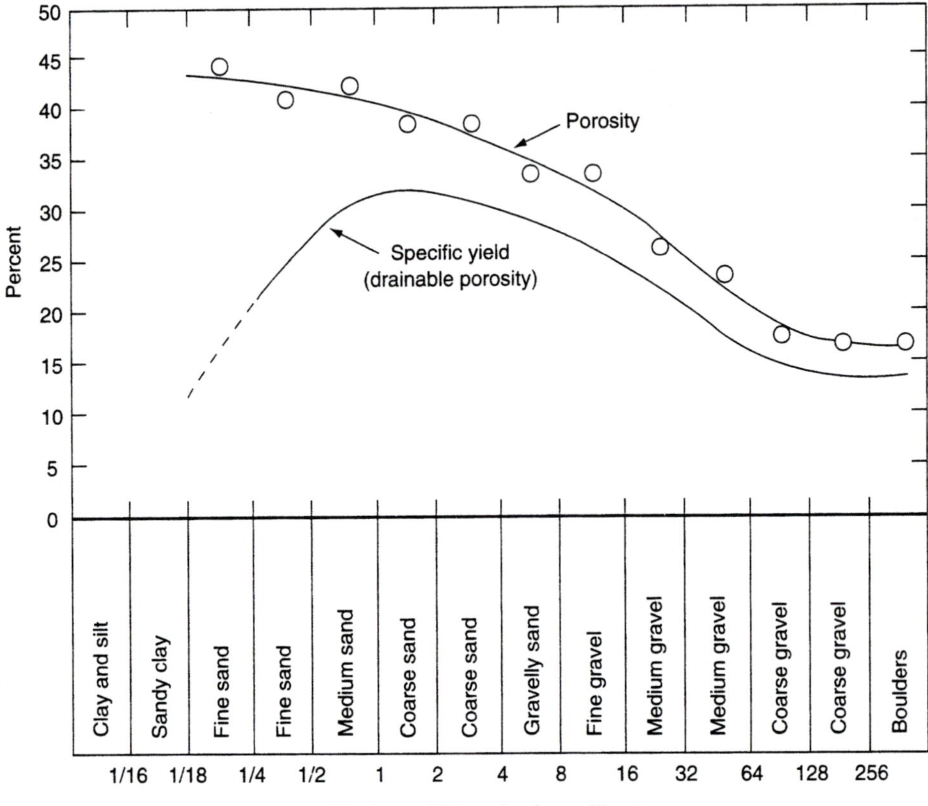

FIGURE 9–6
Chart showing the relationship between porosity and specific yield (similar to effective porosity) for unconsolidated sediments. Note that the specific yield is significantly less than the total porosity for fine grained sediments; for coarse-grained materials, specific yield and porosity are nearly equal (from Barcelona et al., 1985).

techniques because significantly more variability can result from inconsistent purging methods than from all other sources of error combined.

ASTM PS64-96 presents statistical methods that are intended to reduce the frequency of false positive and false negative results during detection monitoring or long-term monitoring programs. The *TEGD* discusses data validation and other statistical techniques that are intended to reduce the number of false positives for a detection-monitoring program. These references will provide additional guidance regarding the statistical performance of a monitoring program.

SUMMARY

In response to physical, chemical and biological processes, groundwater chemistry and movement are constantly in a state of flux. The purpose of the groundwater monitoring program is therefore to detect the spatial and temporal changes for specific parameters and protect sensitive receptors from contaminant migration.

- Through this conceptual discussion of groundwater monitoring, it should be apparent that initial planning stages have far-reaching influence on the success of the monitoring program.
- The hydrogeologist must be devoted to the construction and maintenance of an appropriate Sampling and Analysis Plan. The plan should be reviewed and modified in response to changing site conditions.
- This chapter has introduced you to the widely used guidance documents to prepare, execute, and interpret a groundwater monitoring program.
- The data collected during the groundwater monitoring program are of paramount importance to the reliability of the conceptual hydrogeological model; following the site-specific protocol will result in reduced data variability and a clearer understanding of site conditions.

QUESTIONS FOR REVIEW

1. What are the four types of groundwater monitoring programs? Define and differentiate them.
2. What are the elements of each of the following:
   Health and Safety Plan?
   Sampling and Analysis Plan?
   Quality Assurance and Quality Control Plan?
   Assessment Plan?
3. What is the relationship between an *aquifer* and an *aquitard*? What type of sediments might be representative of each type of unit?
4. In groundwater monitoring programs, it is not uncommon to test for contaminants on the part per billion (ppb) level. For example, the MCL for vinyl chloride in drinking water is 0.002 mg/l (or 2 ppb). To aid you in understanding how minuscule this amount is, let one minute of your day represent "one part." How many years are required for that minute of time to represent one part per billion?
5. What are the primary processes that transport groundwater contaminants in the subsurface? What physical mechanism is responsible for driving each process?
6. What are ten different potential sources of groundwater contamination in the city where you reside?
7. What are the two primary reasons for implementing a groundwater monitoring program at a contaminated site?
8. What is the purpose of groundwater monitoring well development? How does this differ from well purging?
9. What is the purpose of conducting a slug test or pumping test?
10. What are the main sources of variability in a groundwater monitoring program? Describe each.

ACTIVITIES

1. Suppose that it has recently been discovered that the service station that you frequently use had a slow release of gasoline from an underground storage tank system over the last year. Interview the station manager to determine the layout of his underground storage tank and product piping system. Prepare a sketch of the site and formulate an assessment strategy based on the suspected sediment type, potential receptors, depth to groundwater, groundwater flow direction, and characteristics of the fuel released. What type of drilling rig do you propose to use? Why did you choose this method? Describe the monitoring well materials and construction techniques that will be used to assess this site. For what possible contaminants in the groundwater will you be testing?

2. Obtain a copy of the Material Safety Data Sheet (MSDS) for the fuel that you normally purchase from your local service station. Read the MSDS and identify those chemical constituents that are regulated by the EPA under the Safe Drinking Water Act. For each constituent of concern, specify the solubility of the contaminant and rank the relative mobility of each chemical in the subsurface. Note the primary exposure symptoms, pathways, and health hazards associated with the product.

READINGS

American Society for Testing and Materials, ASTM Designation D5092-90, 1990. *Standard Practice for Design and Installation of Ground Water Monitoring Wells in Aquifers*. Scranton, PA: ASTM, 12 pp.

American Society for Testing and Materials, ASTM Designation D4700-91, 1991. *Standard Guide for Soil Sampling from the Vadose Zone*. Scranton, PA: ASTM, 17 pp.

American Society for Testing and Materials, ASTM Designation D4448-85a, 1992, reapproved 1992. *Standard Guide for Sampling Groundwater Monitoring Wells*. Scranton, PA: ASTM, 13 pp.

American Society for Testing and Materials, ASTM Designation D5434-93, 1993. *Standard Guide for Field Logging of Subsurface Explorations of Soil and Rock*, 3 pp.

American Society for Testing and Materials, ASTM Designation D2488-93, 1993. *Standard Practice for Description and Identification of Soils (Visual-Manual Procedure)*. Scranton, PA: ASTM, 11 pp.

American Society for Testing and Materials, ASTM Designation PS64-96, 1996. *Provisional Standard Guide for Developing Appropriate Statistical Approaches for Ground-Water Detection Monitoring Programs*. Scranton, PA: ASTM, 14 pp.

Barcelona, M. J.; Gibb, J. P.; Helfrich, J. A.; and Garske, E. E, 1985. *Practical Guide for Ground-Water Sampling*, USEPA/600/2-85/104. Robert S. Kerr Environmental Research Laboratory. Springfield, MO: U.S. EPA, 169 pp.

Barcelona, M. J.; Wehrmann, H. A.; Schock, M. R.; Sievers, M. E.; and Karny, J. R., 1989. *Sampling Frequency for Ground-Water Quality Monitoring*, EPA/600/S4-89/032. U.S. EPA Environmental Monitoring Systems Laboratory. Washington, DC: U.S. EPA, 6 pp.

Fetter, C. W., 1988. *Applied Hydrogeology*. Columbus, OH: Merrill, 592 pp.

Gates, W. C. B., 1989. "Protection of Ground-Water Monitoring Wells Against Frost Heave." *Bulletin of the Association of Engineering Geologists* 26, no. 2, pp. 241–251.

Lipson, D. S., and Roe, K. J., 1991. "Monitoring Well Placement During the Phase II Site Assessment." In *Proceedings of 1991 Environmental Site Assessments Case Studies and Strategies: The Conference*. Dublin, OH: Water Well Journal Publishing, pp. 287–301.

McLelland S. P., and DeSocio, E. R., 1991. "When Are Monitoring Wells Needed?" In *Proceedings of 1991 Environmental Site Assessments Case Studies and Strategies: The Conference*. Dublin, OH: Water Well Journal Publishing, pp. 279–286.

Puls, R. W., and Barcelona, M. J., December 1995. "Low-Flow (Minimal Drawdown) Ground-Water Sampling Procedures." In *Ground Water Issue*, EPA/540/S-95/504. Washington, DC: U.S. EPA Office of Solid Waste and Emergency Response, 12 pp.

Technical Enforcement Guidance Document, OSWER-9950.1, 1986. *RCRA Ground-Water Monitoring Technical Enforcement Guidance Document*. Washington, DC: U.S. Government Printing Office, 208 pp.

10

Laboratory Methods of Analysis

Sim D. Lessley

Upon completion of this chapter, you will be able to do the following:

- Place the laboratory's contribution to environmental protection in perspective.
- Describe typical laboratory methods of analysis and their relationship to regulations.
- Understand the importance of communicating clearly with a laboratory.

INTRODUCTION

Laboratory analysis follows sampling in the sequence of processes used to develop the information needed to make decisions about the environment. This process begins with the recognition of the requirement, or possible requirement, for environmental action. It progresses through planning, sampling, laboratory analysis, and report preparation. A decision maker uses the information in the report to determine which, if any, action needs to be taken (see Figure 10–1). The important concept to keep in mind is this: Sampling and analysis are performed to provide information that will be used to make a decision.

Decisions and Decision Makers

Decision makers may be line managers or corporate executives, compliance officers, regulators, or judges. The decisions to be made are many and varied. Some examples follow:

- Is a plant's effluent below permitted discharge levels? If not, what action should be taken?
- How far has the liquid leaking from a derailed tank car spread? Is it necessary to evacuate people from the neighborhood?

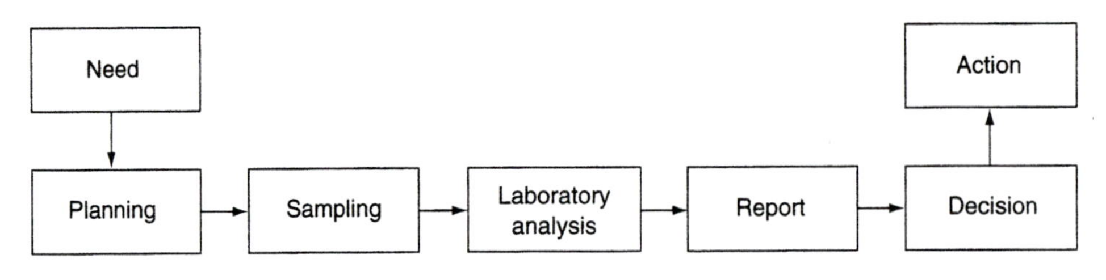

FIGURE 10–1
Information used for decisions.

- Which areas of a decommissioned military installation can be turned over for immediate public use? Which areas must be cleaned up before public use?
- How much cleanup will a leaking service station tank require?
- Is the contamination at a particular site the result of criminal negligence?

The Importance of Protocols

In general, the people who perform the sampling and analysis work do not meet the decision makers, for several reasons. With the exception of emergencies, the decision usually follows sampling and analysis by months, or even years. Frequently, the individuals who perform the different processes in the sequence work for different organizations, and the different processes are carried out at various locations. For example, the need for an environmental survey at a decommissioned military installation may be determined by a Department of Defense administrator in Washington, D.C. The planning may be contracted out to one engineering company, and the sampling may be performed by a different firm from a different city. A laboratory in yet another city may be subcontracted by the sampling group. Usually the engineering company that handles the sampling also prepares the report. The decision makers in such a scenario may be a group with representatives from the Department of Defense, the state in which the installation is located, and the EPA Regional Office.

Because of this level of complexity, *protocols* (standardized procedures) assume great importance in environmental projects. Some protocols are mandated by the regulations applying to the situation. Others are defined as part of the planning process. *Laboratory methods* are protocols that clearly detail how analytical procedures are to be performed. Laboratory methods must be specified and followed in order for the laboratory to produce data suitable for the decision makers' uses.

THE LABORATORY'S CONTRIBUTION TO ENVIRONMENTAL PROTECTION

The Role of the Laboratory

Referring again to the flow diagram in Figure 10–1, you can see that the laboratory's role is to receive samples, perform analyses, and provide data for a report to be used by a decision maker. Actually, two types of laboratories contribute to environmental protection in different ways. The first type researches new analytical techniques, develops analytical methods, and monitors their performance. The second type of laboratory applies established methods to the analysis of samples, typically in large numbers. This type of laboratory tends to modify methods only to the extent required by the sample matrix.

The Research Laboratory

The first type of laboratory, the *research laboratory*, seeks to develop and refine analytical methods. Its goal is to add new analytical methods, while evaluating and improving existing ones. Usually the research laboratory is part of a nonprofit organization, typically a governmental agency. Examples of such laboratories are the EPA Environmental Measurement and Support Laboratories (EMSLs), the Army Corps of Engineers' laboratories, and the Department of Energy National Laboratories, as well as some college and university laboratories. Developing new analytical methods, evaluating method performance, and recommending improvements to methods require a widely skilled staff with enough time to develop creative approaches and investigate them. Generally, the research laboratory is not equipped to analyze large numbers of samples.

The Produ\ction Laboratory

The *production laboratory* is organized to analyze large numbers of samples cost effectively, according to established methods. Staff skills are more focused, and a production laboratory has a higher proportion of technicians than a research laboratory. Usually the production laboratory is a business operating for profit. As such, it is structured to process many samples in a short time at optimum unit costs. Examples of production laboratories include full-service commercial laboratories, and "niche laboratories" that specialize in a particular type of analysis. Some nonprofit agency laboratories, such as those in public health departments, also operate as production laboratories.

The remainder of this chapter focus on production laboratories. The terms *laboratory* and *analytical laboratory* will be used to refer to production laboratories, because most environmental samples today are analyzed by that type of laboratory.

Laboratory Services

The core group of services to be expected from a production environmental analytical laboratory includes the following:

- Maintenance of laboratory accreditation and compliance with its requirements. Laboratory accreditation requirements cover staff qualifications, quality assurance/quality control procedures, documentation and records retention, and disposal of samples and laboratory waste, as well as certification to perform specific methods.
- Prompt receipt and log-in of samples, with feedback to the client if sample integrity has been compromised. Common examples of compromised sample integrity are broken or leaking sample containers, temperature in the shipping chest out of range, and incorrect addition of chemical preservatives.
- Appropriate storage and subsampling to ensure that the portion taken for analysis truly represents the sample as received.

- Analysis of samples within the required holding times. Sample holding times usually are specified in the regulations that apply to the project.
- Analysis of the samples according to the methods previously agreed upon.
- An analytical report that has been certified to be accurate and that notes anything unusual about the analysis of the samples.

Depending on its size and client requirements, a laboratory may offer additional supplies and services such as these:

- Preservatives, clean containers, and shipping chests for sample collection.
- Trip blanks.
- Chain-of-custody forms and evidence tape.
- Assistance with sample collection or sample pickup.
- Assistance in writing bids and proposals.
- Review of project plans.
- Expert witnesses.

Data sets produced by different laboratories using the same sampling method should be comparable. Otherwise, analytical laboratories are as varied as the personalities of their founders. Rarely do the style and blend of services offered by one laboratory exactly match those of another. Therefore, it is wise to become familiar with a laboratory so you may understand exactly how that laboratory will fit with the other components of your project. A visit to the laboratory, combined with a clear discussion of project requirements, is the best way to reach that level of understanding.

THE RELATIONSHIP OF LABORATORY METHODS TO REGULATIONS

The Development of Regulations—A Review

Local, state, and federal agencies often respond to unwanted incidents by developing regulations intended to prevent such incidents from occurring again. The environmental regulations with which we live and work were developed in response to unwanted incidents from four categories:

1. Industrial accidents.
2. Occupational safety.
3. Consumer protection.
4. The environmental protection movement.

This development process is described in detail in Chapter 1, "History of Environmental Regulations," in *Environmental Regulations Overview*, edited by Neal K. Ostler and John T. Nielsen, 1996 (Volume 2 in this series).

As regulations to prevent unwanted incidents were developed, it became clear that the results of chemical analyses were necessary to administer some of them. This was especially true of those regulations intended to control long-term problems such as increases in human cancer rates, heightened mortality of endangered species, declines in water quality, and increases in atmospheric pollution. Analytical results generally were not necessary to administer regulations intended to prevent catastrophic incidents such as explosions, fires, and wrecks. As the demand for the results of chemical analysis came into focus, a related demand began to grow as well—the requirement for analytical measurements at lower and lower concentration levels. Research suggested that many of the long-term problems resulted from the cumulative effects of exposure to very small amounts of toxic materials over an extended time. Accordingly, to effectively administer the regulations, it was necessary to accurately measure very small amounts of toxic materials. As new environmental regulations were being developed in the 1960s and 1970s, cases in which no standard analytical methods were available to accomplish the newly required measurements appeared, sometimes because the analytical technology to reach the low levels simply did not exist.

The Development of Analytical Methods

Analytical methods supporting of water purification, commerce, and agriculture were developed and gathered into manuals beginning in the late 1800s. In 1895, members of the American Public Health Association (APHA) recognized the need for standard methods in the bacteriological examination of water. A committee was appointed to draw up procedures. These were submitted in 1897 and found wide acceptance. In 1899, the APHA appointed a committee on standard methods of water analysis, responsible for developing standard procedures for all methods involved in the analysis of water. The committee's report, published in 1905, was titled "Standard Methods of Water Analysis." It constituted the first edition of what is commonly called *Standard Methods*. The nineteenth edition of *Standard Methods for the Examination of Water and Wastewater* was published in 1995. The American Society for Testing and Materials (ASTM) began to publish analytical methods in 1898. The Association of Official Agricultural Chemists was founded in 1884, and began to publish copyrighted methods in 1916. It later became known as the Association of Official Analytical Chemists (AOAC).

As environmental legislation in the 1960s and 1970s was passed, the resulting regulatory programs looked first to existing collections of analytical methods, such as *Standard Methods* and those of the ASTM and the AOAC. In some cases the existing methods were incorporated by reference; in others, they were rewritten into a format more consistent with the regulations. The newly formed EPA undertook development of analytical methods to fill the "method gaps" described previously. As time passed, many of the analytical methods went through revisions that improved their performance and extended them to include more compounds. Three important collections of analytical methods, which are associated with the major pieces of environmental legislation, are summarized in Table 10–1.

The *List of Lists*

Although the preceding three method compendiums encompass most of the commonly used analytical methods, a number of others have been developed to address specific requirements of a variety of regulations. By 1990 there were so many regulatory lists, substances, and analytical methods that the EPA recognized the need to clarify which should be used to determine the compounds of interest under the different regulations. It published the *List of Lists* in 1990 and updated it in 1991. (*List of Lists—A Catalog of Analytes and Methods*, U.S. EPA Office of Water [WH-552],

TABLE 10–1
Three important collections of analytical methods.

Methods for Chemical Analysis of Water and Wastes
Methods for Chemical Analysis of Water and Wastes, EPA-600/4-79-020, 1979, was developed to support the Clean Water Act (CWA) and Safe Drinking Water Act (SDWA) regulations. Updates were released in 1983, 1991, and 1993. (Many of these methods are rewritten procedures from *Standard Methods*).
Test Methods for Evaluating Solid Waste (SW-846)
The third edition of *Test Methods for Evaluating Solid Waste*, 3rd ed. (*SW-846*) was developed to support Resource Conservation Recovery Act (RCRA) regulations. This collection of methods has been improved upon and expanded steadily since it first appeared. Update IIB to the third edition was promulgated in January 1995. A draft of Update III was released for review and comment in late 1995. It should be noted that these methods apply to the analysis of both water and solids, in spite of the compendium title.
Contract Laboratory Program (CLP) Methods
Chemical analyses to support environmental actions taken under the Comprehensive Environmental Response, Compensation, and Liability Act (CERCLA) are performed by commercial laboratories under contract to the EPA. This effort is the Contract Laboratory Program (CLP). CLP methods are fewer in number when compared to the other collections, but they are the most aggressively revised and improved. It is particularly important to specify which version of a CLP method is to be used, because the methods are revised so frequently. Generally, it is best to use the current version.

21W-4005, August 1991.) The *List of Lists* is organized by chemical substance. For each chemical substance, the applicable method(s) and regulation(s) are given. Many substances are regulated under more than one program and determined by more than one method.

The EPA subsequently released the "Environmental Monitoring Methods Index" (EMMI). The EMMI is available on electronic media, allowing searches of 50 EPA regulatory lists, 2,600 substances, and 926 analytical methods. Both the *List of Lists* and the EMMI may be obtained from the National Technical Information Service (NTIS). The NTIS telephone number is (703) 487-4650.

METHODS OF ANALYSIS—WET CHEMISTRY

The wet chemistry methods of analysis, whose descriptions and applications are described in this section, encompass four areas:

1. Spectrophotometry.
2. Electrode methods.

3. Gravimetric methods.
4. Titrimetric methods.

These methods were developed originally for the analysis of drinking water and wastewater. They have been used for decades, and are very well established.

Spectrophotometry

Absorption *spectrophotometry* measures the interaction of electromagnetic radiation with matter. It involves measuring the intensity of two radiation beams as a function of wavelength. Figure 10–2 illustrates the absorption of radiation. To measure absorption, you must determine the intensity of the incident radiation (P_0) and the transmitted radiation (P_t). The efficiency of the absorption of radiation as a function of wavelength is illustrated in Figure 10–3, which presents an absorbance spectrum. Materials absorb radiation more efficiently at certain wavelengths. The differences in absorbance result from differences in atomic and molecular structure.

The absorbance of electromagnetic radiation by atoms and molecules is quantized; that is, the absorption (or emission) of electromagnetic radiation results in a change in the energy state (energy content) of the molecule or atom. Changes in the energy state of an atom or a molecule occur in steps outlined in Equation 10–1.

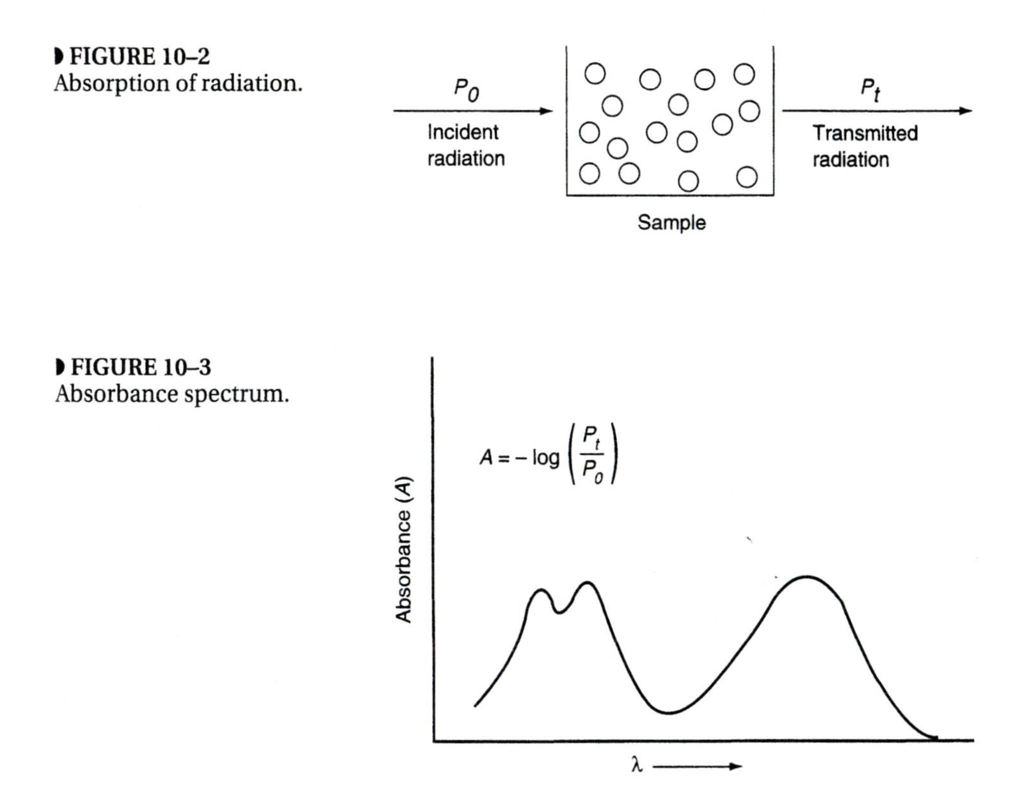

FIGURE 10–2
Absorption of radiation.

FIGURE 10–3
Absorbance spectrum.

Equation 10–1

$$\Delta E = hc/\lambda$$

Where

ΔE = the change in energy state.

h = Planck's constant

c = the speed of light.

λ = the wavelength of the electromagnetic radiation.

Ultraviolet and visible electromagnetic radiation will cause a transition of electrons between energy levels in atoms and molecules. Infrared radiation causes a transition of molecules between vibrational and rotational energy states.

Figure 10–4 is a schematic of a single-beam spectrophotometer. Electromagnetic radiation from a light source is focused on a wavelength selector. The wavelength selector (a diffraction grating, a prism, or a filter) is used to isolate a wavelength at which the material of interest absorbs radiation. The radiation passes through the sample cell and is focused on a photomultiplier tube that measures its intensity. Referring back to Figure 10–2, the incident radiation (P_0) is measured with the sample cell empty, and the transmitted radiation (P_t) is measured with the material of interest in the sample cell.

The absorbance of radiation is described by Equation 10–2, the *Beer-Lambert law.*

Equation 10–2

$$A = \varepsilon bc$$

Where

$$A = -log_{10}\left(\frac{P_t}{P_0}\right)$$

ε = molar absorptivity (the efficiency of absorption of radiation).

b = path length (length of the path through the sample cell).

c = concentration of the material in solution.

The spectrophotometer and the Beer-Lambert law can be used to measure the concentration of an environmental pollutant. Figure 10–5 illustrates a plot of absorbance versus concentration. The absorbances of standards of known concentration (the black dots) are measured and plotted against concentration. Subsequently, the

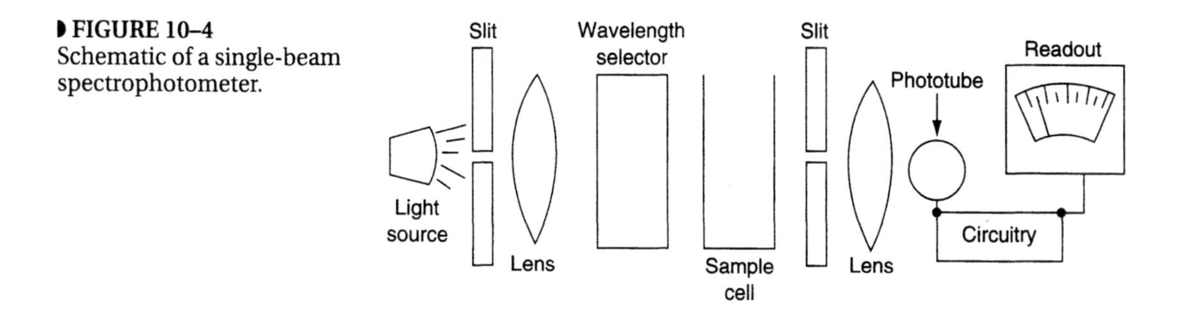

FIGURE 10–4
Schematic of a single-beam spectrophotometer.

absorbance of a sample (the open dot) can be measured and the concentration calculated from the line's equation.

Many analysis methods found in *Standard Methods* and in the EPA's *Methods for Chemical Analysis of Water and Wastes* are spectrophotometric. In *Methods for Chemical Analysis of Water and Wastes*, spectrophotometry is used to determine chloride (325.2), cyanide (335.3), ammonia (350.1), nitrate (353.2), phosphate (365.4), silica (370.1), and sulfate (375.2), to name a few. In each case the sample is prepared for analysis and then compared to standards of known concentration using a spectrophotometer set at a wavelength appropriate for that particular compound. Spectrophotometric methods usually are sensitive and comparatively inexpensive. Their weakness lies in a susceptibility to interferences. An *interference* is a material that is not the compound of interest but affects the absorbance of the compound of interest. For example, phenol interferes with the spectrophotometric analysis of formaldehyde. Many spectrophotometric methods have been automated, which increases their reliability and reduces cost.

Electrode Methods

The determination of pH and fluoride are examples of *electrode methods* (Methods 150.1 and 340.2 in *Methods for Chemical Analysis of Water and Wastes*). The potential (voltage) from an electrode sensitive to the material of interest is measured against a standard potential. The measured potential is related to the concentration of the material of interest by a mathematical expression called the *Nernst equation*. For routine analyses, the Nernst equation is not used to calculate results. Instead, a meter with special built-in circuitry "solves" the equation and presents the concentration of the material of interest as a direct readout.

Gravimetric Analyses

Examples of *gravimetric analyses* are the Total Residue Method (160.3) and the Oil and Grease Method (413.1) from the EPA's *Methods for Chemical Analysis of Water and Wastes*. Gravimetric methods require weighing small amounts of the material of

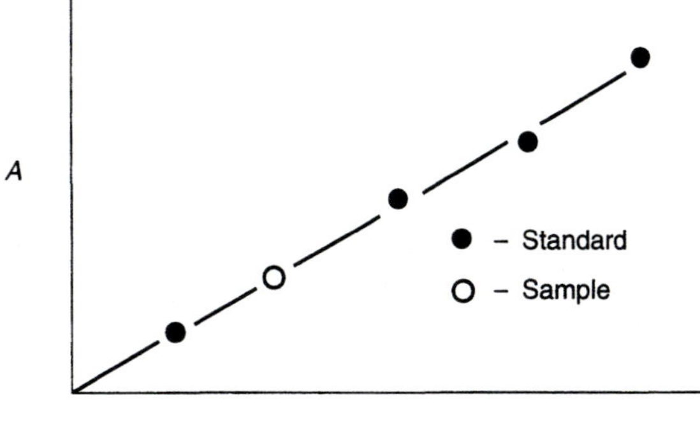

FIGURE 10–5 Application for quantitative analysis.

interest on a sensitive laboratory balance. The balances used for this purpose read out to 0.1 or 0.01 mg.

Titrimetric Methods

Examples of *titrimetric methods* are the Determinations of Sulfide and Sulfite (Method 376.1 and Method 377.1 in *Methods for Chemical Analysis of Water and Wastes*). Titrimetric methods of analysis require dispensing accurately measured volumes of reagents of known concentration. At the "end point" of a titrimetric analysis, the added reagent exactly balances the concentration of the material of interest. In Method 376.1, 1 mL of 0.250 normal standard iodine solution will react with 0.4 mg of sulfide present in the sample. The amount of sulfide in the sample can be calculated from the volume of standard iodine solution used in the titration. The end point of the titration typically is determined by the appearance or disappearance of a color, or by an electrode similar to those used in the electrode methods described previously.

METHODS OF ANALYSIS—ATOMIC ABSORPTION AND ATOMIC EMISSION

Atomic absorption (AA) and atomic emission (AE) are specializations of the spectrophotometry technique. Atoms are characterized by different electronic states with differing energy levels. Atoms may *absorb* electromagnetic radiation (light), resulting in a change from a lower to a higher energy level. Atoms also may *emit* electromagnetic radiation (light), resulting in a change from a higher to a lower energy level. The wavelength of electromagnetic radiation that may effect an electronic transition is very specific for that particular transition. For example, the wavelength (λ) of the warm yellow cast characteristic of sodium vapor lights is 589.0 nanometers (nm). The extent to which radiation of a given wavelength is absorbed or emitted by a particular type of atom may be used for quantitation. AA and AE analytical techniques can be used to determine one particular type of atom in a mixture of other types of atoms by selecting a wavelength of light unique to that particular atom. The intensity of electromagnetic radiation absorbed or emitted can be related directly to the amount of the atoms of interest present in the sample.

Although in theory all elements in the periodic table could be determined by AA or AE, some practical limitations apply. The elements at the top right-hand corner of the periodic table (Figure 10–6) are not commonly analyzed by AA or AE because their unique wavelengths are outside the range of commonly available instruments. The elements at the bottom of the periodic table, the lathanide series and the actinide series, are not often analyzed by AA or AE because the wavelength of each is very similar to that of the other elements in the series. The elements amenable to determination by AA or AE are found to the left and in the middle of the periodic table. Examples are lithium (Li), beryllium (Be), sodium (Na), magnesium (Mg), potassium (K) through selenium (Se), and rubidium (Rb) through antimony (Sb).

A successful analysis by AA or AE depends on converting the element of interest to its atomic state. Only isolated atoms in the gaseous state, bonded to nothing else, are able to absorb or emit the unique wavelengths on which the analysis depends.

IA	IIA	IIIB	IVB	VB	VIB	VIIB	VIIIB			IB	IIB	IIIA	IVA	VA	VIA	VIIA	VIIIA
1 H 1.0079																	2 He 4.0026
3 Li 6.941	4 Be 9.012											5 B 10.81	6 C 12.01	7 N 14.01	8 O 16.00	9 F 19.00	10 Ne 20.18
11 Na 22.99	12 Mg 24.31											13 Al 26.98	14 Si 28.09	15 P 30.97	16 S 32.97	17 Cl 35.45	18 Ar 39.95
19 K 39.10	20 Ca 40.08	21 Sc 44.96	22 Ti 47.88	23 V 50.94	24 Cr 52.00	25 Mn 54.94	26 Fe 55.85	27 Co 58.93	28 Ni 58.69	29 Cu 63.54	30 Zn 65.39	31 Ga 69.72	32 Ge 72.61	33 As 74.92	34 Se 78.96	35 Br 79.90	36 Kr 83.80
37 Rb 85.47	38 Sr 87.62	39 Y 88.91	40 Zr 91.22	41 Nb 92.91	42 Mo 95.94	43 Tc 98.91	44 Ru 101.1	45 Rh 102.9	46 Pd 106.4	47 Ag 107.9	48 Cd 112.4	49 In 114.8	50 Sn 118.7	51 Sb 121.8	52 Te 127.6	53 I 126.9	54 Xe 131.3
55 Cs 132.9	56 Ba 137.3	57 La to Lu 71	72 Hf 178.5	73 Ta 180.9	74 W 183.9	75 Re 186.2	76 Os 190.2	77 Ir 192.2	78 Pt 195.1	79 Au 197.0	80 Hg 200.6	81 Tl 204.4	82 Pb 207.2	83 Bi 209.0	84 Po 209	85 At 210	86 Rn 222
87 Fr 223	88 Ra 226.0	89 Ac to Lr 103	104 Rf 261.1	105 Ha 262.1	106 Sg 263.1	107 Nr 262.1	108 Hs	109 Mt									

Lanthanide Series

57 La 138.9	58 Ce 140.1	59 Pr 140.9	60 Nd 144.2	61 Pm 144.9	62 Sm 150.4	63 Eu 152.0	64 Gd 157.3	65 Tb 158.9	66 Dy 162.5	67 Ho 164.9	68 Er 167.3	69 Tm 168.9	70 Yb 173.0	71 Lu 175.0

Actinide Series

89 Ac 227.0	90 Th 232.0	91 Pa 231.0	92 U 238.0	93 Np 237.0	94 Pu 244.1	95 Am 243.1	96 Cm 247.1	97 Bk 247.1	98 Cf 251.1	99 Es 252.1	100 Fm 257.1	101 Md 258.1	102 No 259.1	103 Lr 260.1

FIGURE 10–6
Periodic chart of the elements.

The sample preparation procedures and the "front end" of the AA and AE instruments are designed to break all chemical bonds and deliver isolated atoms into the instrument's optical path for measurement.

Atomic Absorption Spectroscopy

Water, soil, or waste samples are leached and dissolved in strong acid solutions. Hydrochloric acid (HCl) and nitric acid (HNO_3) are used most often. The purpose of this sample preparation step is to liberate the atoms of analytical interest from the other materials in the sample matrix and to break as many chemical bonds to the atoms of interest as possible. All of the bonds of some elements, such as sodium (Na), are broken by treatment with strong acid. Others, such as aluminum (Al), remain partially bonded to oxygen even after treatment with strong acid. The atomic absorption spectrophotometer is designed to remove the acid solution from the element of interest and put it into the gaseous atomic state. It also is designed to break the remaining bonds of refractory elements such as aluminum (Al). This vaporization and final bond breaking are accomplished by high temperatures.

A *flame atomic absorption* (*FAA*) instrument uses a gas burner to reach the desired temperature. Commonly used gas mixtures are air–natural gas, air–acetylene, and nitrous oxide–acetylene. Flame temperatures can reach 1,700–2,000°C. Figure 10–7 is a diagram of the FAA measurement process. A very small stream of the sample that has been dissolved in acid is drawn up into the nebulizer and is converted into an aerosol spray. The oxidizer and the fuel are blended with the aerosol in the mixing chamber and swept up into the burner. The light source used in FAA is a lamp with a filament made from the element of analytical interest. Atoms of the same element as this filament absorb the unique light wavelengths in proportion to their concentration. The wavelength of analytical interest is selected using the monochromator, which is typically a diffraction grading. Finally, the light intensity is measured by a photomultiplier tube detector.

FAA is a well-developed and reliable analytical technique. The instrumentation is comparatively inexpensive. However, inadequate sensitivity has proven to be a problem in the analysis of some elements. For example, the limits of detection for arsenic and lead by FAA are approximately 500 ppb and 50 ppb, respectively. Environmental decision makers typically require detection limits of 1 to 5 ppb.

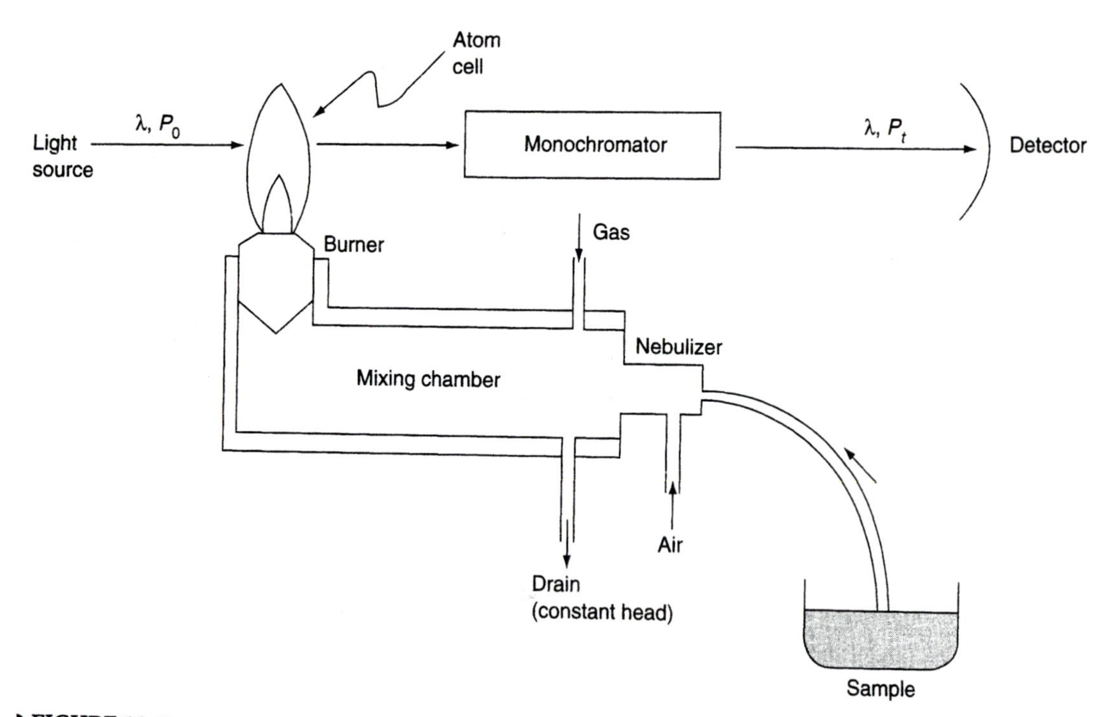

FIGURE 10–7
Flame atomic absorption spectroscopy.

Graphite Furnace Atomic Absorption Spectroscopy

The *graphite furnace atomic absorption* (*GFAA*) technique is able to achieve detection limits in the 1 to 5 ppb range. To accomplish this, the flame is replaced by a small carbon tube heated by electrical resistance. The graphite furnace operates at a higher temperature (2,000–2,800°C) than the flame and requires less sample. However, it is much more prone to interferences and requires a higher degree of operator skill.

Cold Vapor Atomic Absorption Spectroscopy

Mercury is determined by a unique procedure called *cold vapor atomic absorption (CVAA)*. In this analytical technique, the sample is digested in a strong acid solution to solubilize the mercury. The mercury ions are subsequently reduced to mercury vapor by means of a tin-containing reagent. The absorbance of the mercury atoms is measured at a wavelength of 184.9 nm. A flame AA can be converted easily to a cold vapor mercury analyzer simply by replacing the burner head with a flow cell for the mercury vapor.

Atomic Emission Spectroscopy

The *atomic emission spectroscopy (AES)* technique relies on the same electronic transitions in free atoms that were described in atomic absorption methods. Instead of measuring absorbed radiation, the AES technique measures the intensity of emitted radiation. A diagram of a flame emission spectrophotometer is found in Figure 10–8. Sample preparation and introduction are the same as the flame atomic absorption

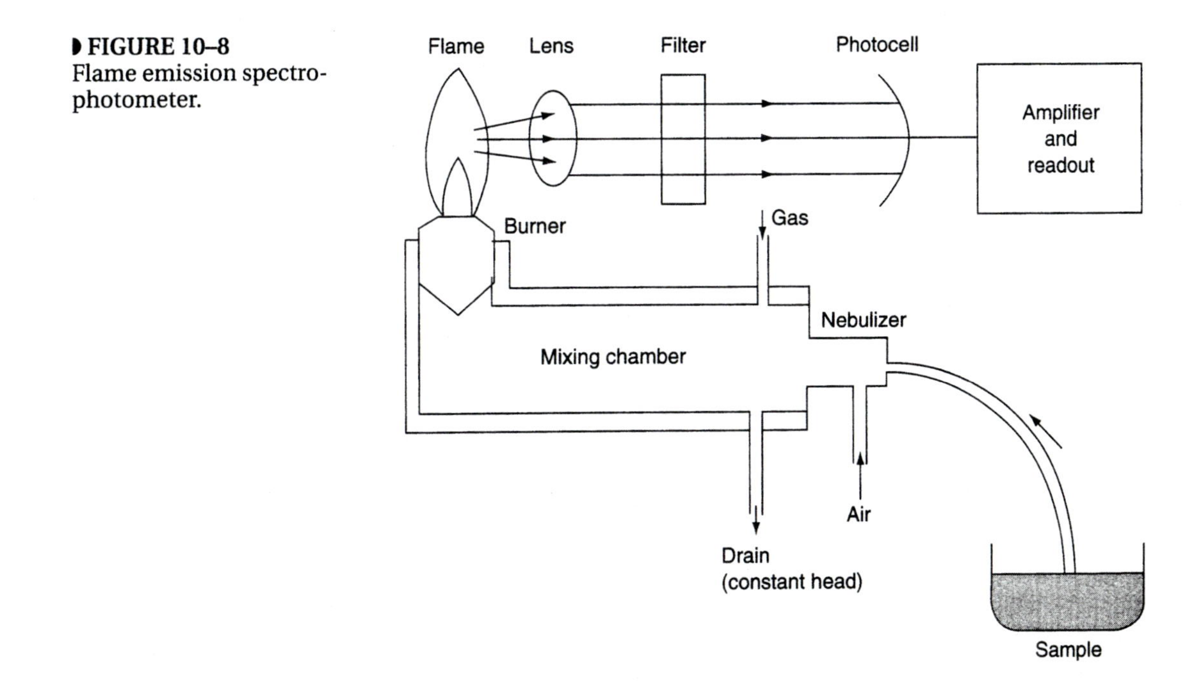

FIGURE 10–8 Flame emission spectrophotometer.

approach. The difference between the two techniques lies in the measurement of emitted radiation rather than absorbed radiation. AES offers the possibility of measuring several elements at one time. This advantage has been optimized in the design of the inductively coupled plasma–atomic emission spectrometer.

Inductively Coupled Plasma–Atomic Emission Spectometry

The *inductively coupled plasma–atomic emission spectrometer* (*ICP–AES* or *ICP*) replaces the flame with an argon plasma that operates at about 10,000°C. The ICP has excellent sensitivity in the part-per-billion range. The plasma's high temperature results in few chemical interferences. Instrument vendors have focused significant research and development efforts on ICP, resulting in instruments that can measure up to 50 elements at one time.

The ICP has become a workhorse technique for determining toxic metals in environmental samples. Method 6010A from the third edition of *SW-846* and the EPA–CLP ICP method are most frequently used for this purpose. A typical combination of methods for assessing metal contamination at an environmental site consists of:

1. Up to twenty-six metals analyzed by ICP, according to Method 6010A from *SW-846*.
2. Arsenic, lead, and selenium analyzed by GFAA, according to Methods 7060, 7421, and 7740, respectively, from *SW-846*.
3. Mercury analyzed by CVAA, according to Methods 7470 (liquids) and 7471 (solids) from *SW-846*.

METHODS OF ANALYSIS—GAS CHROMATOGRAPHY

Gas chromatography (*GC*) is applied to the analysis of samples containing volatile compounds of interest. The volatile compounds may be gases at standard conditions, or they may liquids or solids that can be converted to vapor by heating. Most compounds in this category are organic molecules with a molecular weight of less than 1,000. Nonvolatile materials not suitable for gas chromatography include high-molecular-weight (more than 1,000) polymers such as wood or plastics, highly polar organic compounds, and most inorganic compounds.

Sample Preparation

Environmental samples of water, soil, or waste usually are prepared for GC analysis by extraction. Water samples are shaken up with a solvent that does not mix with water, such as methylene chloride. When the layers separate, the solvent layer is collected over a drying agent (typically sodium sulfate). Solid samples are exposed to the same solvents for several hours or, alternatively, exposed to the solvents for a shorter time in an ultrasonic device. Again, the solvent is collected over the drying agent. After the solvent has evaporated to a small volume, usually 1–10 mL, it is ready for analysis by GC.

Especially volatile organic compounds (VOCs) such as chloroform, carbon tetrachloride, or benzene cannot be prepared for analysis by extraction because they would be lost during evaporation. VOCs are introduced into the GC by a *purge-and-trap* (P&T) process. The water sample, or a water slurry of the solid sample, is placed in a tightly sealed vessel that allows helium gas to bubble through it. This is the purge step. The stream of helium sweeps the volatile components out of the sample and into a tube containing charcoal and a proprietary polymer called Tenax®. Silica gel also is present to remove the water vapor. This is the trapping step. Subsequently, the tube is heated rapidly to drive the trapped volatile components into the gas chromatograph for analysis.

The Gas Chromatograph

Figure 10–9 represents the basic components of a gas chromatographic system. A chemically inert carrier gas such as helium or nitrogen flows continuously through the system. The sample is introduced into the continuously flowing gas stream by means of the injector. The sample components are separated in the column, which is maintained at a specified temperature by the column oven. As the separated components in turn elute from the end of the column, the detector measures the amount of each. The signal from the detector is amplified, digitized, and routed to a data system that stores and performs calculations with the data. It also produces a pictorial record of the separation, called a *chromatogram*.

The Chromatogram

The chromatogram contains two types of information: *Qualitative* information answers the question "What is present? and *quantitative* information answers the question "How much is present?" Figure 10–10 contains a simple chromatogram illustrating the concept of qualitative analysis. The chromatogram records the signal

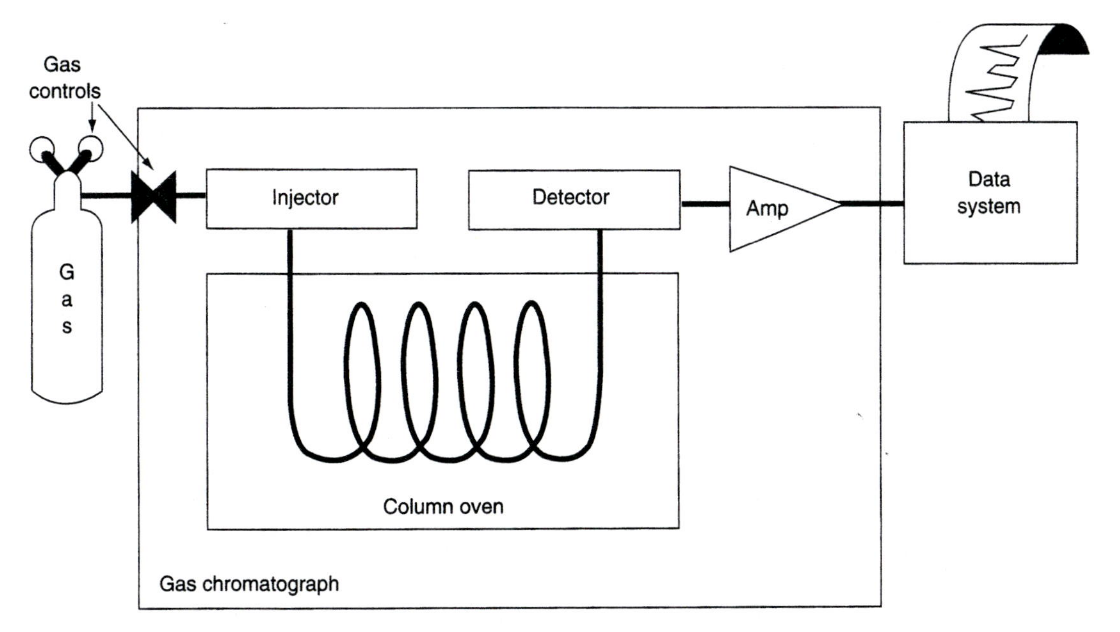

FIGURE 10–9
Basic components of a gas chromatographic system.

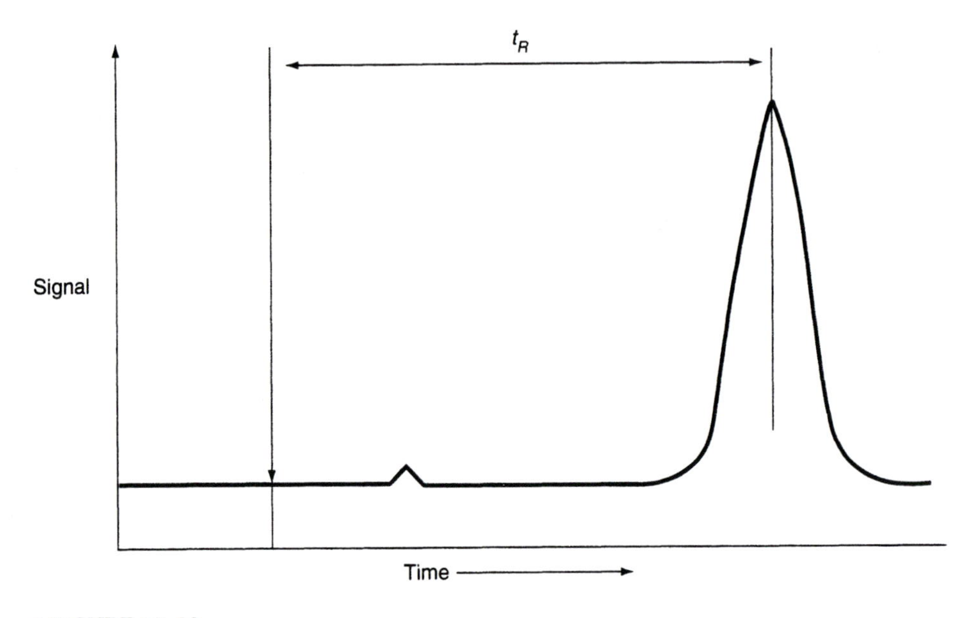

FIGURE 10–10
A simple chromatogram illustrating a qualitative analysis.

from the detector on the *vertical* axis and the time elapsed between injection and the detector response on the *horizontal* axis. This time is called the *retention time* (t_R). It represents the time that the substance corresponding to the peak is retained in the chromatographic column under the specified conditions.

Figure 10–11 illustrates a chromatogram with three components separated. Each component in the standard mixture has a corresponding retention time (t_R). The (t_R) of methylethyl ketone (MEK) is illustrated. Suppose that a sample of unknown composition were analyzed on the gas chromatograph under exactly the same set of conditions. The second chromatogram in Figure 10–11 illustrates the analysis of the unknown mixture. One component was separated. The fact that the retention time (t_R) of the unknown component matches the retention time of MEK obtained under exactly the same conditions leads to the conclusion that the unknown sample contains MEK. Qualitative analysis, the identification of materials in samples, is achieved in gas chromatography by matching retention times with standards of known composition. This qualitative analysis procedure actually applies to all forms of chromatography.

Qualitative analysis based on retention time is always open to question because of the possibility that a different compound could have the same retention time under the conditions of the analytical method. The identification of a compound can be considered more reliable by analyzing the sample on a different column under a different set of conditions. If the retention times of the standard and the component in the sample match again, the identification (qualitative analysis) is said to be confirmed. This process is called *second-column confirmation.*

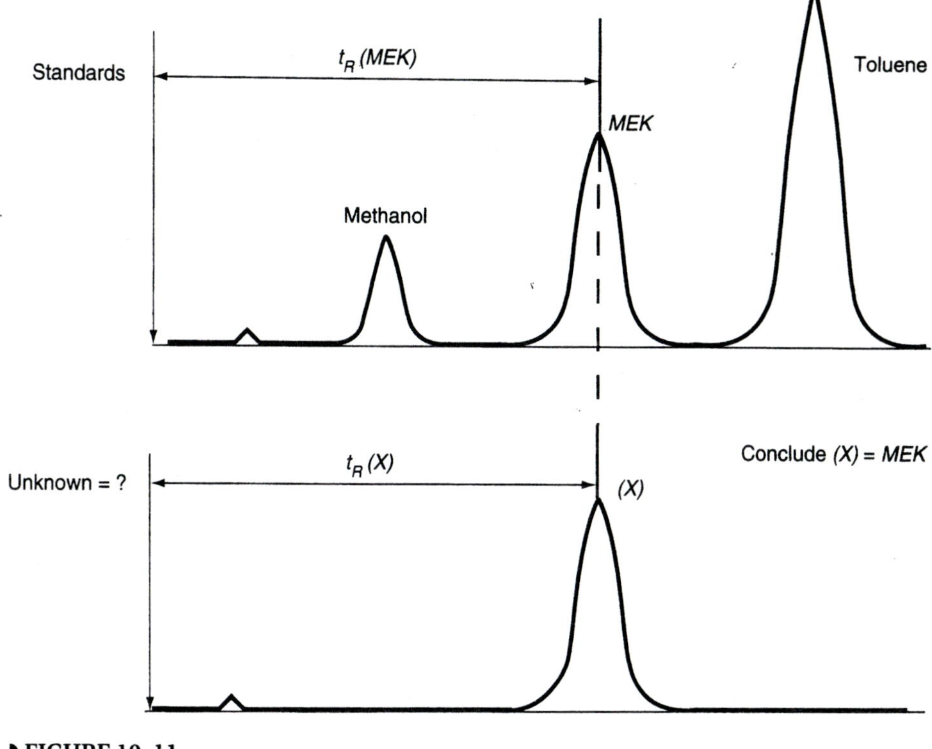

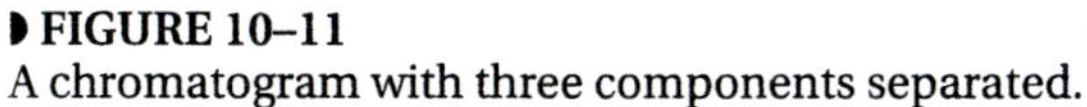

FIGURE 10–11
A chromatogram with three components separated.

Figure 10–12 illustrates quantitative analysis, the determination of how much of each component is present. This may be accomplished by measuring the peak height of a component and comparing it to the peak heights in the chromatograms of standards obtained under exactly the same conditions. Quantitative analysis also may be accomplished by measuring the peak area and comparing it to those from chromatograms of standards obtained under exactly the same conditions. In the early days of chromatography, the qualitative and quantitative analyses were accomplished by making measurements on the chromatogram. A modern gas chromatograph uses the data system to acquire retention times, peak heights, and peak areas. A chromatogram still is produced for the operator's use in evaluating the analysis.

The GC Column

A packed GC column is illustrated in Figure 10–13. The column packing consists of small particles (80 to 120 mesh) coated with a liquid phase. Crushed fire brick and calcined diatomaceous earth are common solid supports. A variety of high-molecular-weight liquids with different functional groups bonded to them has been used as liquid phases. Silicone oils are among the most popular. A packed column usually is

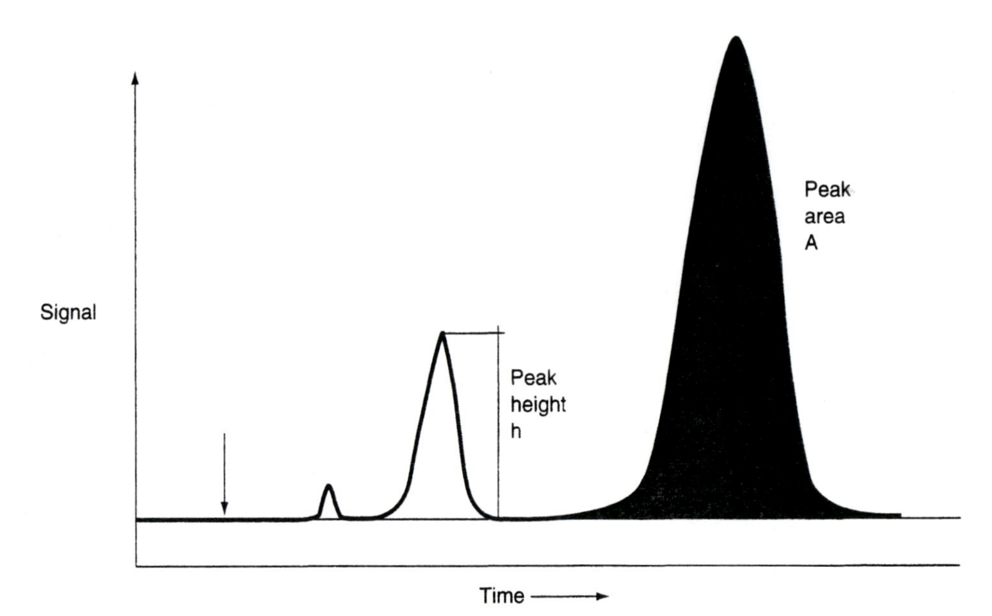

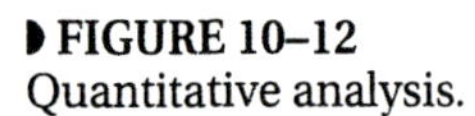
FIGURE 10–12
Quantitative analysis.

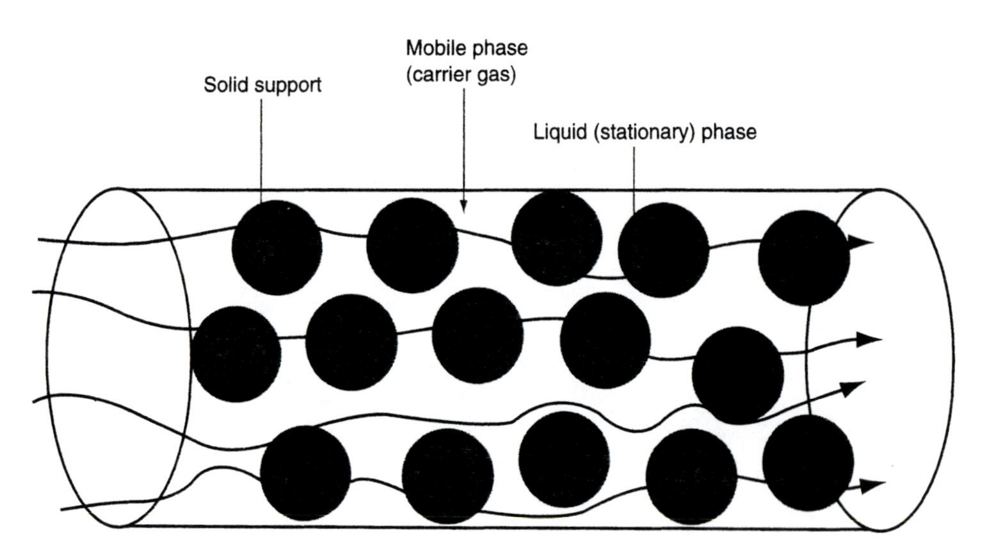

FIGURE 10–13
GC column.

made of glass or stainless steel, and typically is 3–20 feet long, with a diameter of 1/8–1/4 inch. When a packed column is being used in a GC, the inert carrier gas (helium or nitrogen) flows around and through the tiny spaces between the small particles of packing material. The components of a mixture injected into the GC column are separated because they have different affinities for the liquid phase coated on the solid

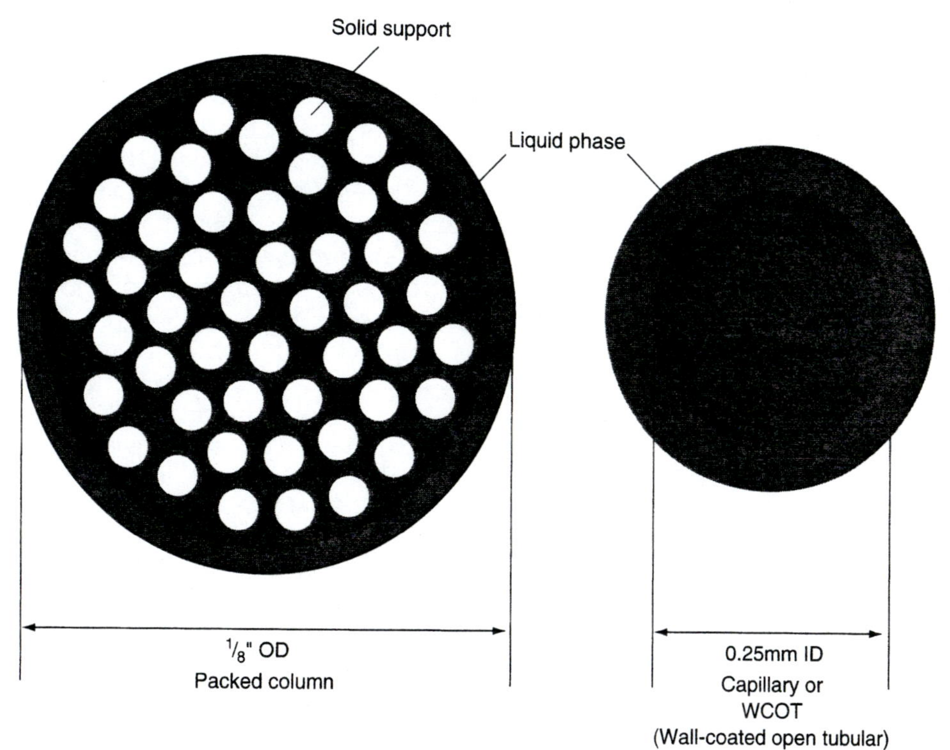

FIGURE 10–14
GC column types.

support. Components with less affinity for the packing come out of the column first with shorter retention times. Those with greater affinity for the liquid phase on the packing proceed through the column more slowly and come out later with longer retention times.

During the last twenty years, the packed column has been displaced from the production laboratory by a capillary or wall-coated open tubular (WCOT) column. Capillary columns are made of fused silica with a polyimide coating on the outside to minimize breakage. Capillary columns are much smaller in diameter (0.25mm–1.0mm ID) and much longer (30m–100m) than packed columns. Instead of being coated on small particles, the liquid phase in a capillary column is a 1-micron-thick coating that is chemically bonded to the inside walls. Figure 10–14 illustrates the differences between a packed column and a capillary column. The resolution of a capillary column is superior to that of a packed column; that is, the capillary column does a much better job of separating components in a mixture. This superior resolution can be applied to separating components of environmental interest from complex interferences. The superior resolution also can be "traded off" for a shorter run time through the gas chromatograph; that is, the capillary column is able to achieve equivalent separation in significantly less time than a packed column.

Detectors

The detector in a gas chromatograph is fitted tightly to the end of the column. It responds to the separated components as they elute from the column, converting each to an electrical signal that is proportional to the amount of each component present. A variety of GC detectors has been developed. The choice of a detector is based on a trade-off among several attributes. Four of the most important are sensitivity, response, linearity, and stability. Usually, environmental decision makers are interested in measurements of very low concentrations of environmental pollutants. Accordingly, the sensitivity of a GC detector is important. Sensitivity commonly is presented as a minimum detectable quantity. Some GC detectors respond to a broad range of compounds, while others respond quite selectively to a narrow range. A detector's *linearity* is a measure of the range over which its response can be mathematically modeled by the equation for a straight line ($Y = mx + b$). Linearity typically is expressed in powers of ten. The stability of a detector has to do with how long it can operate without needing to be recalibrated with standards.

In the process of developing standard analytical methods, the trade-offs between sensitivity, response, linearity, and stability have been worked out. The standard analytical methods usually specify one (or occasionally two) detector types to be used for the analysis.

The Thermal Conductivity Detector (TCD)

The *thermal conductivity detector* is based on the Wheatstone bridge circuit. In the TCD, two of the resistors in the bridge circuit are heated filaments. One of the filaments is continuously bathed in a flow of carrier gas; The other filament is surrounded by the effluent from the column. As the separated components flow past the second filament, its ability to dissipate heat is changed, which results in an imbalance in the current flow in the bridge circuit. This difference in current flow is fed through an amplifier into the GC data system. The following list summarizes the TCD's characteristics:

Sensitivity—fair.
Response—universal.
Linearity—good.
Stability—good.

Field gas chromatographs equipped with TCDs are used to analyze environmental samples on site.

The Flame Ionization Detector (FID)

The *flame ionization detector* utilizes a small air–hydrogen flame that burns between two electrodes. As the separated components from the GC column flow into the flame, the carbon-containing compounds are burned, producing charged ions. As the charged ions move between the electrodes, they generate a small current that, again, is amplified and fed into the GC data system. The characteristics of the FID are summarized as follows:

Sensitivity—good.
Response—selective, only organic molecules.
Linearity—very good.
Stability—excellent.

Method 8015 from *SW-846* has been modified for the analysis of samples contaminated with hydrocarbon fuels (gasoline, diesel, kerosene, etc.). It uses a GC with an FID (GC/FID).

The Nitrogen Phosphorus Detector (NPD)

The *nitrogen phosphorus detector* has a heated rubidium bead positioned between two electrodes. A plasmalike condition is established on the surface of the bead. As the separated components from the column flow across the bead, compounds containing nitrogen or phosphorus cause a current to flow between the two electrodes. The current is amplified and fed into the GC data system. The properties of the NPD are summarized here:

Sensitivity—very good.
Response—very selective for nitrogen (N) and phosphorus (P).
Linearity—fair.
Stability—fair to poor.

Method 8070 from *SW-846* calls for a GC equipped with an NPD for the analysis of samples that contain nitrosamines.

The Flame Photometric Detector (FPD)

The *flame photometric detector* is a miniaturized version of the flame emission spectrophotometer previously described in the discussion of AA/AE methods of analysis. This highly selective detector typically is applied to the analysis of sulfur- or phosphorus-containing compounds. As the separated components containing sulfur or phosphorus flow from the column into the detector, they are burned in a small flame. The characteristic sulfur or phosphorus wavelengths are selected by a filter in the light path. The resulting signal from a photomultiplier tube is amplified and fed into the GC data system. The characteristics of the flame photometric detector are summarized as follows:

Sensitivity—good.
Response—very selective for sulfur or phosphorus.
Linearity—poor.
Stability—fair to good.

Method 8141A from *SW-846* uses a GC/FPD for the analysis of samples containing organophosphorus pesticides such as malathion and parathion.

The Electron Capture Detector (ECD)

The *electron capture detector* contains a small amount of foil made from a radioisotope of nickel (Ni-63). Ni-63 emits beta particles (electrons) as it decays. As the separated components from the column flow through the ECD, those with an affinity for electrons acquire a negative charge. The movement of these negatively charged compounds between two electrodes generates a small current that can be amplified and fed into the GC data system. The ECD is one of the most sensitive and most selective GC detectors available. It responds to compounds that contain halogens (chlorine, fluorine, bromine, and iodine), as well as compounds containing nitro groups. The ECD is very susceptible to interferences and requires an unusually high degree of operator skill. The characteristics of the electron capture detector are as follows:

Sensitivity—excellent.
Response—very selective.
Linearity—poor.
Stability—fair to good.

Organochlorine pesticides and polychlorinated biphenyls (PCBs) are determined according to *SW-846* Method 8080A using a GC/ECD.

GC Data Systems

GC data systems have made tremendous improvements in recent years—a result of the availability of personal computers. As computing power has increased and prices have dropped, sophisticated data-handling programs have become widely available in production laboratories. This enhanced capability has two distinct advantages: First, the determination of retention times and peak areas or heights is done much more consistently; second, the entire data-reduction and reporting process is accomplished much more efficiently.

METHODS OF ANALYSIS—GAS CHROMATOGRAPHY/MASS SPECTROMETRY

The *gas chromatograph/mass spectrometer (GC/MS)* is a powerful analytical tool that combines the superior separation capability of a gas chromatograph with the remarkable capability for qualitative analysis offered by the mass spectrometer. The advantages of gas chromatography as a separation technique were described in a foregoing section. The weakness of gas chromatography lies in the margin of uncertainty that surrounds identifying compounds by their retention time (t_R). The fact that the retention time of a peak in the chromatogram of an unknown sample matches the retention time of a standard (trichloroethene, for example) obtained under the same conditions strongly suggests that trichloroethene is present. However, there are other organic compounds that could have the same retention time as trichloroethene under that particular set of experimental conditions. Such compounds could mistakenly be identified as trichloroethene, giving a "false positive" result. Many analytical methods undertake to strengthen the GC's qualitative analysis results by requiring confirmation of identity on a second column of different composition. If the second chromatogram from a different column taken under different conditions confirms the identification by matching the retention time of the standard, the qualitative analysis becomes much stronger. Still, some doubt always will remain regarding the identification of a compound based on retention times alone. The mass spectrometer overcomes this difficulty through "fingerprinting" each chromatographic peak by determining its mass spectrum. With the exception of a few lower molecular-weight compounds, the mass spectrum provides a sure identification of compounds because of the uniqueness of each mass spectrum.

The Mass Spectrometer

The components of a mass spectrometer are illustrated in Figure 10–15. To function, the ion source, the mass analyzer, and the ion collection system must be maintained under a high vacuum. The purpose of the inlet system is to transfer the separated components into the ion source as they elute from the column, without degrading the high vacuum required for proper operation. As a component from the GC column passes through the inlet system and enters the ion source, it is converted into charged particles by a beam of high-energy electrons. Most of the charged particles are ionic fragments of the parent molecule. In some situations the parent molecule itself is converted to a parent ion through the loss of an electron. The mass analyzer then sorts (separates) the ions according to their mass/charge (m/z) ratio. Finally, the ion collection system produces a *mass spectrum,* which is the record of relative numbers of different kinds of ions from the fragmentation of the parent compound. The fragmentation pattern (mass spectrum) is characteristic of the parent molecule and the energy of the electron beam. The fragmentation pattern usually is sufficient to identify the parent molecule. If the parent ion can be observed, the molecular weight can be determined as well. The data-handling system can be programmed to add the ions corresponding to each component that elutes from the column. The resulting trace is called a *total ion chromatogram,* or *reconstructed ion chromatogram,* which looks very similar to the output of any other GC detector.

Figure 10–16 illustrates the process of *electron impact (EI) ionization* that occurs in the ion source. Though other types of mass analyzers are available, the quadrupole mass analyzer is most frequently used in production analytical laboratories. It is illustrated in Figure 10–17. The quadrupole mass analyzer may be viewed as a type of "filter" that allows ions of different m/z through to the ion collector as the voltages applied to the quadrupoles change.

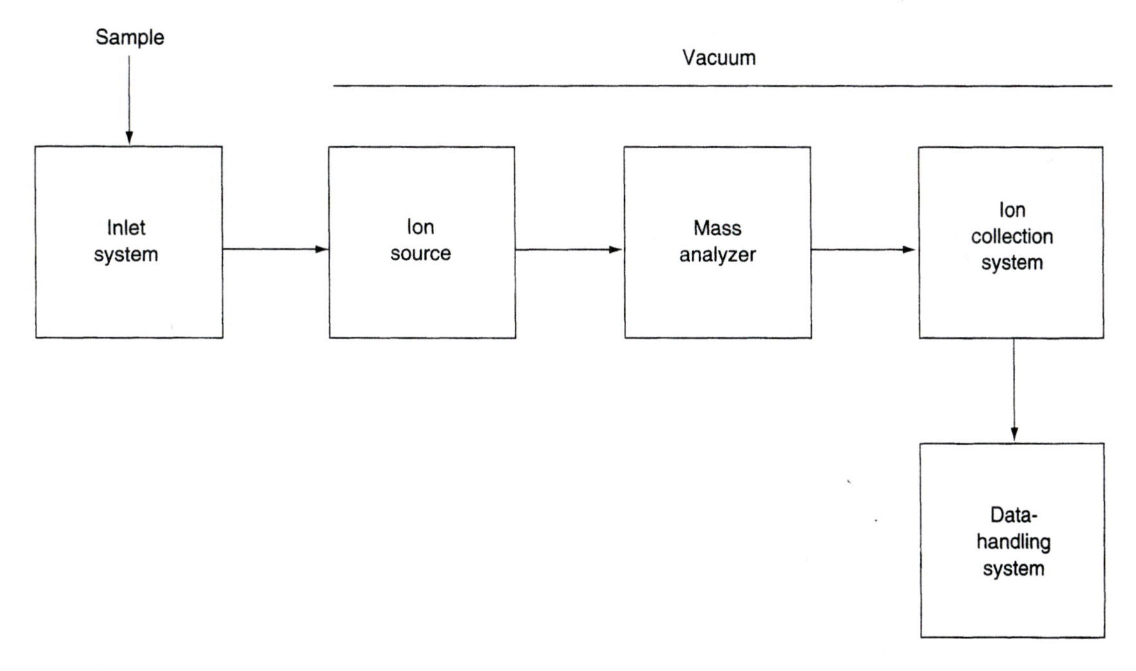

FIGURE 10–15
Components of a mass spectrometer.

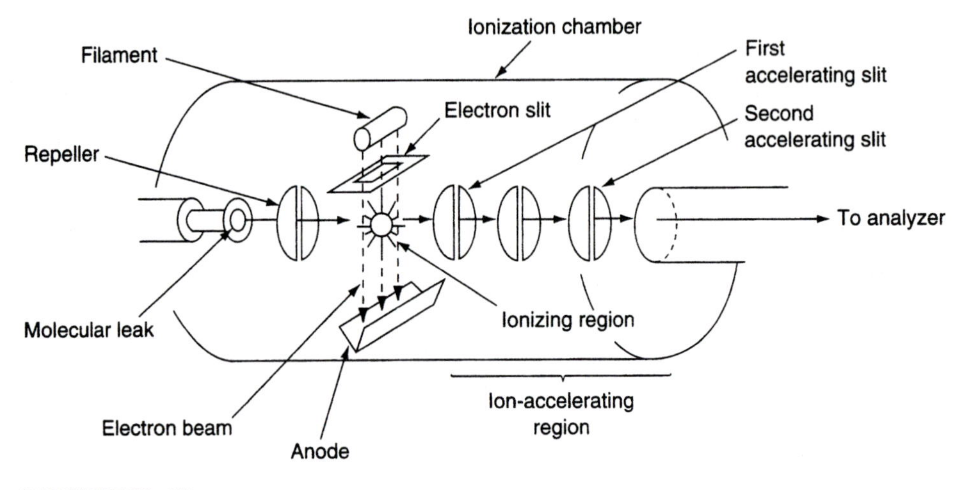

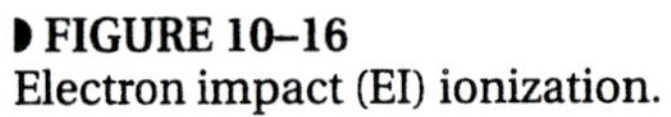
FIGURE 10–16
Electron impact (EI) ionization.

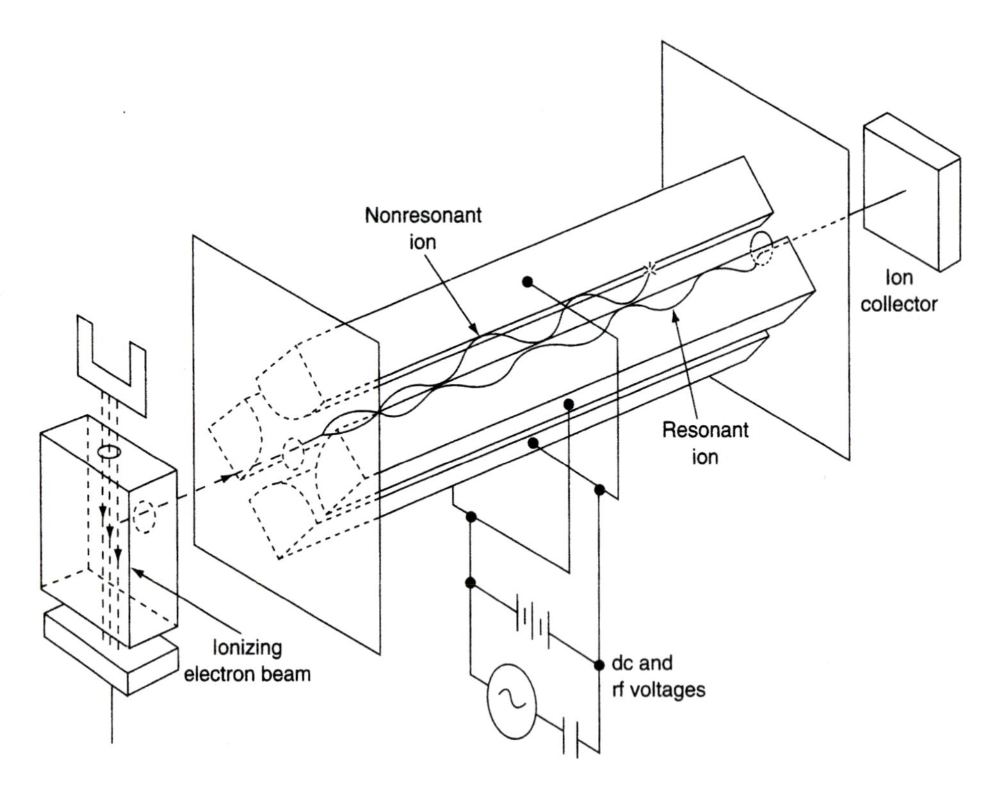

FIGURE 10–17
Quadrupole mass analyzer.

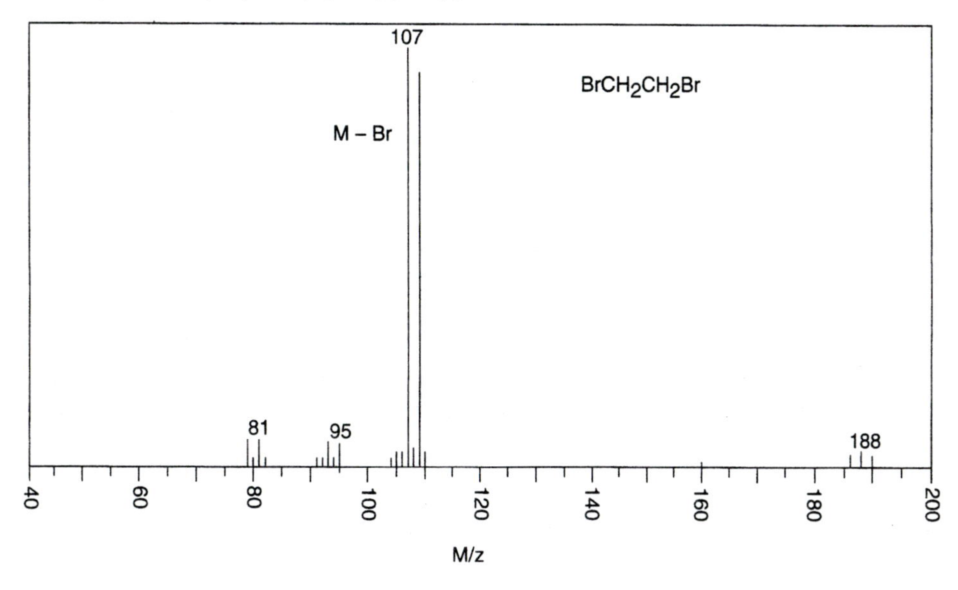

FIGURE 10–18
Mass spectrum of ethylene dibromide.

The mass spectrum of ethylene dibromide is presented in Figure 10–18. The mass spectrum of 1, 3, 7, 8-tetrachlorodibenzodioxin is presented in Figure 10–19. As the mass spectra of the separated components are generated, in turn they are stored in the data-handling system. A computer program compares each mass spectrum from the sample analysis to a database of mass spectra that has been prepared by NIST. Presently, there are more than 70,000 compounds in the NIST library. The data system prints out each component's mass spectrum and the mass spectra of the three compounds that most closely match it. The final identification is made by a trained analyst after reviewing the data. The GC/MS also produces a quantitative analysis report based on the comparison of one or two ions from each component to known amounts of standards.

Applications of GC/MS

GC/MS ideally is suited to screen environmental samples for a large variety of contaminants. Method 8270B, the semivolatile GC/MS method from *Test Methods for Evaluating Solid Waste (SW-846),* lists 257 potential target compounds. Usually, the instrument is calibrated with a standard for each of about 100 compounds. Method 8260A, the GC/MS method for volatile compounds, includes 97 compounds in the target list; usually about 60 compounds are determined in actual practice. In addition, nontarget compounds can be tentatively identified using the NIST library.

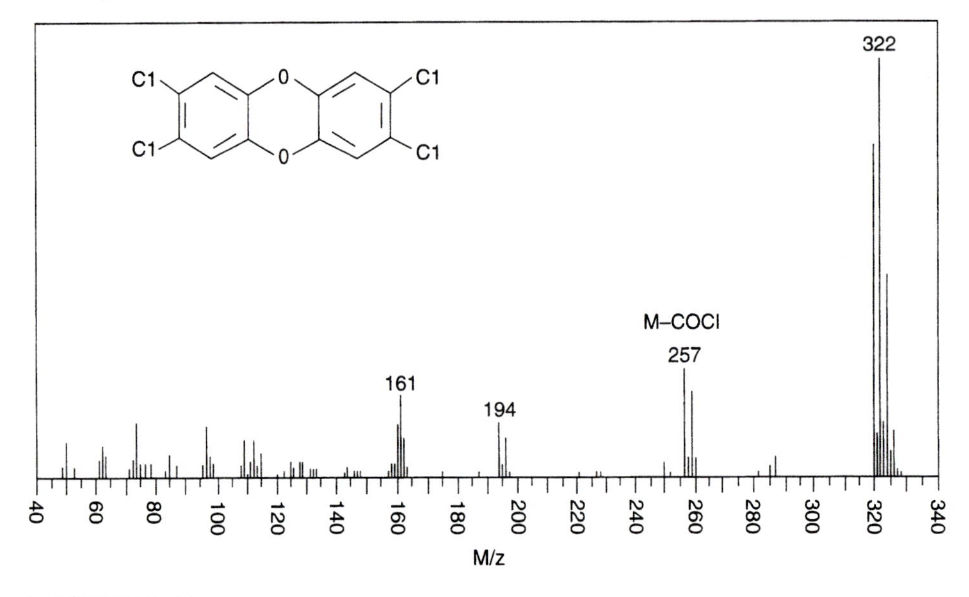

FIGURE 10–19
Mass spectrum of 1, 3, 7, 8 tetrachlorodibenzodioxin.

It should be noted that some GC detectors are more sensitive than the mass spectrometer. Therefore, after site assessment is complete, some compounds may be monitored by GC rather than by GC/MS.

METHODS OF ANALYSIS—LIQUID CHROMATOGRAPHY

A variety of analytical techniques is included in the general category of liquid chromatography. The same general principles described in relationship to gas chromatography apply to liquid chromatography as well. Specifically, liquid chromatographs use columns and achieve separation by partitioning between a mobile phase and a stationary phase. The difference is that the mobile phase, instead of being an inert gas such as helium or nitrogen, is instead a liquid. Figure 10–20 contains an overview of liquid chromatography. Looking at the left side of the diagram, water-soluble environmental pollutants typically have molecular weights under 2,000. The technique most frequently used for such compounds is ion-exchange chromatography. The water-insoluble compounds (right side of the diagram) of environmental interest also usually have a molecular weight less than 2,000. These compounds are analyzed by adsorption and partition chromatography, commonly called *high-performance liquid chromatography (HPLC)*. Compounds of environmental concern usually do not have molecular weights over 2,000. Therefore, *gel permeation chromatography (GPC)*, also called *size-exclusion chromatography*, is not

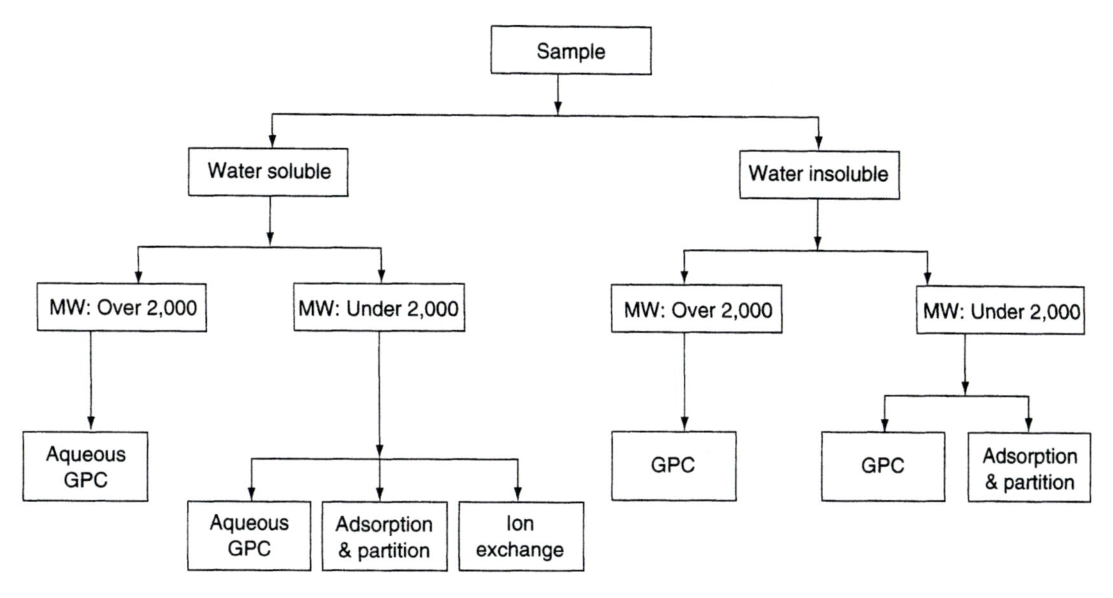

FIGURE 10–20
Liquid chromatography—an overview.

usually performed in a production analytical laboratory. There is one exception: Method 8080(A) from *Test Methods for Evaluating Solid Waste (SW-846)* allows for GPC to remove higher molecular-weight interferences from samples that are to be analyzed for pesticides.

Ion-Exchange Chromatography

EPA Method 300.0 is an *ion-exchange chromatography* method, which was promulgated for the analysis of drinking water samples under the Safe Drinking Water Act (SDWA). Some environmental programs permit the use of Method 300.0 for wastewater and groundwater samples, but others do not. The method employs a column filled with an ion-exchange resin and uses a dilute carbonate/bicarbonate buffer solution as the mobile phase (eluent). Typical ions determined by this method include fluoride, chloride, nitrite, nitrate, sulfate, and phosphate.

Ion-exchange chromatography offers a cost-effective way to determine several ions at once. It has good sensitivity and is readily automated. Ion-exchange chromatography is subject to the uncertainty surrounding identification of compounds that is common to all chromatographic techniques. In some situations, heavy metal ions present in the sample matrix can degrade the method's sensitivity for certain ions, such as phosphate.

High-Performance Liquid Chromatography (HPLC)

Figure 10–21 contains a schematic of a typical high-performance liquid chromatography (HPLC) system. The mobile phase generally is a mixture of water with a polar organic solvent such as methanol or acetonitrile. HPLC columns are normally 10

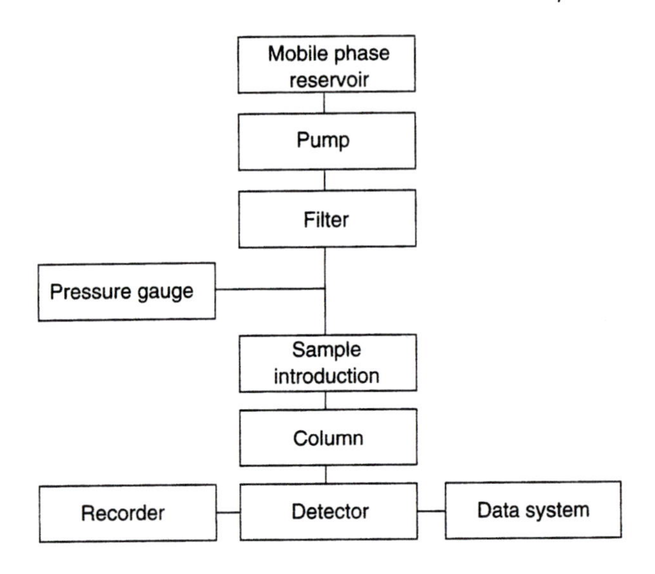

FIGURE 10–21
High-performance liquid chromatography (HPLC) system.

inches long by ¼ inch in diameter. They are packed very carefully with small, uniform particles. The result is a column with excellent separation capability; however, pressure in the range of 6,000 to 10,000 psi is required to force the mobile phase through the column. As a result, care must be taken to ensure that particles from the sample are filtered out so that they do not damage the pump or clog the column.

The UV–visible detector is most commonly used for environmental samples. It is an absorbance-type detector, similar in principle to the one described in the wet chemistry section. Modern HPLC systems also have benefited from advances in data-handling systems. Sophisticated software packages similar to those used in GC systems determine retention times, peak areas, and peak heights, as well as calculate concentrations of components of interest based on concentrations of standards.

Applications of HPLC

Method 8310 from *Test Methods for Evaluating Solid Waste (SW-846)* is used to determine polynuclear aromatic hydrocarbons (PAHs) in environmental samples. PAHs usually appear in residue from burning. They also are associated with asphalt and tar. The field samples are extracted with methylene chloride. The methylene chloride extract is reduced in volume and the methylene chloride is exchanged for acetonitrile. The acetonitrile extracts are analyzed by HPLC using a UV–visible detector, as well as a fluorescence detector for confirmation.

Method 8330, also from *SW-846*, was developed under contract to the United States Army for the determination of explosives and their breakdown products. For example, the method determines 2,4,6-trinitrotoluene (TNT), as well as four related dinitrotoluenes. The dinitrotoluenes result from the breakdown of TNT in the environment. As military bases are decommissioned and returned to public use, sampling and analysis for explosives is an important part of those site assessments. Method 8330 is one of the few HPLC methods that includes a second-column confirmation.

LABORATORY QUALITY ASSURANCE/QUALITY CONTROL (QA/QC)

Samples are collected and laboratory measurements are performed to generate information for environmental decision makers. The goal of such decisions is to improve the quality of human health and the environment. In addition to these issues, the laboratory data users' decisions also frequently involve the expenditures of large amounts of money and the assessment of penalties and fines. Because of the importance of their decisions, environmental decision makers have an additional requirement for *assurance* that the laboratory measurements are of *adequate quality*. Five descriptors commonly are used to assess laboratory quality. The first letter of each descriptor, taken in order, makes up the acronym PARCC. The following are the five descriptors:

- **Precision**—The degree of mutual agreement characteristic of independent measurements as the result of repeated application of the process under specified conditions.
- **Accuracy**—The degree of agreement of a measured value with the true or expected value of the quantity of concern.
- **Representativeness**—The degree to which the data accurately and precisely represent a characteristic of a population parameter, variation of a property, a process characteristic, or an operational condition.
- **Completeness**—A measure of the amount of data obtained from a measurement process compared to the amount that was expected to be obtained under the conditions of measurement.
- **Comparability**—The confidence with which one data set can be compared to another.

John K. Taylor, formerly with NIST, has published a very useful book on the subject of laboratory quality assurance (*Quality Assurance of Chemical Measurements*, Lewis Publishers, 1987). In his book, Taylor offers three definitions related to quality assurance:

- **Quality assurance**—A system of activities whose purpose is to provide to the producer or user of a product or service the assurance that it meets defined standards of quality with a stated confidence.
- **Quality control**—The overall system of activities whose purpose is to control the quality of a product or service so that it meets the needs of users. The aim is to provide quality that is satisfactory, adequate, dependable, and economic.
- **Quality assessment**—The overall system of activities whose purpose is to provide assurance that the overall quality control job is being done effectively. It involves a continuing evaluation of the products produced and of the performance of the production system.

Analytical results obtained directly from the analytical method usually are inadequate for environmental decision makers' use. In addition to the analytical result, the decision maker needs to understand the precision, accuracy, representativeness, completeness, and comparability of that result. To accomplish that, the measurement process must be "wrapped" with a quality assurance system. A functioning laboratory quality assurance system has two components: quality control and quality assessment. A well-managed environmental analytical laboratory will document its approach to quality assurance in a quality assurance program plan (QAPP). The QAPP is a document that frequently references other documents on file in the laboratory, such as analytical methods and standard operating procedures (SOPs). In some situations, a laboratory's QAPP will be supplemented for a particular project with a quality assurance project plan (QAPjP).

Quality Control

Revisiting the definition, "quality control is the overall system of activities whose purpose is to control the quality of a product or service so that it meets the needs of users. The aim is to provide quality that is satisfactory, adequate, dependable, and economic." Laboratory quality control techniques include all of the actions taken with regard to measurement processes in an effort to ensure that requirements are met. A laboratory quality control system will address the following components: personnel, facilities, equipment, written procedures, inspection, validation, and documentation.

Personnel

Qualified staff members are the most important component of a quality control system. It is becoming more common for laboratories to maintain files that document the qualifications of each staff member. A typical qualification file contains a position description specifying the requirements and responsibilities of the position that the person holds in the laboratory. If education is required, it is documented by a copy of the degree or certificate. Experience is documented in the file by a résumé. The dates and contents of specific training sessions are listed. A professional attitude on the part of staff members is of great importance as well. Although difficult to document in a file, it can be evaluated easily during a laboratory visit.

Facilities

One of the primary goals of a laboratory facility is contamination control. Methylene chloride, an extraction solvent commonly used to prepare organic samples for analysis, also is one of the target analytes in analytical methods for volatile solvents. Floor plans, construction, traffic patterns, and ventilation must be arranged to ensure that such processes do not contaminate each other. Steady supplies of pure laboratory gases and "clean" electrical power are important. Some analytical methods are sensitive to changes in temperature, so it is necessary to control the temperature in the laboratory. Temperature control of sample storage units also is necessary.

Equipment

A production analytical laboratory needs to have adequate amounts of equipment for two reasons. First, the laboratory must have enough equipment to perform the work it has agreed to do. Second, it must have reserve equipment to use in the event of an equipment failure. Equipment appropriate for the analytical methods must be used. Frequently, in a production laboratory, this includes consideration of auto samplers and computer control. Each piece of laboratory equipment needs to have preventive maintenance performed according to a schedule. The equipment must be correctly calibrated. Most laboratory instruments are calibrated by using them to analyze standards of known composition: the calibration should be verified with a standard every time the instrument is used. When the instrument fails calibration, it should be recalibrated. Timely replacement of marginally functional or outmoded equipment is necessary, especially in view of the fact that expectations of laboratories continue to rise.

Written Procedures

There is an increasing demand for laboratories to maintain files of written procedures. A small laboratory with an experienced staff that performs a limited number of analytical methods may be able simply to refer to the methods in the reference manual. Most environmental analytical laboratories perform a large number of different analytical methods using a mix of experienced and inexperienced staff members. This has resulted in a requirement that laboratories write *standard operating procedures* (SOPs). An SOP starts with a reference method rewritten to explain exactly how that laboratory will implement the reference method. Different analysts in a laboratory may interpret the reference method differently. The goal of the SOP is to eliminate differences in the way a result is obtained in the laboratory.

Inspection

The goal of inspection is to prevent the release of questionable laboratory results. Samples are inspected upon receipt at the laboratory for broken containers, correct storage temperature, and adequate chemical preservation. When a laboratory analyst has completed a report, a reviewer will check adherence to procedure, transcriptions, and calculations. Before the report is sent out, another review may be made to ensure that all supporting data requested actually go out with the analytical report. Frequently, checklists are used in the laboratory to ensure that inspection steps are consistently and thoroughly carried out.

Validation

Validation consists of steps taken to ensure that a laboratory procedure actually performs as expected. Method-detection limits are determined at specified time intervals to validate instrument sensitivity. Samples of known composition are processed through analytical methods to validate the accuracy and precision of the process. Computer programs, particularly data-transfer and data-reduction programs, are checked to validate error-free performance.

Documentation

The final record of sampling and analysis is the documentation. This is very vital because the sample is used in the analysis, the residue is discarded, and laboratory analysts move on to other assignments and after a period of time cannot recollect the details of a particular analysis. As noted in the introduction to this chapter, the environmental decision maker may consider a laboratory result months or years after it was obtained. Under these circumstances, the usability of analytical data depends directly on the quality of the documentation. Documentation is so important that SOPs and checklists are used to ensure that it is done correctly. Laboratory data other than instrument outputs typically are recorded in ink. Any errors are crossed out with a single line, corrected, dated, and initialed. Each set of data is signed by the analyst and then is reviewed and countersigned by the reviewer. After completion, laboratory documents typically are maintained in secure, fire-proof storage for three to five years, unless otherwise specified in a contract.

Quality Assessment

Quality assessment, as stated previously, "is the overall system of activities whose purpose is to provide assurance that quality control is being done effectively. It involves a continuing evaluation of the products produced and of the performance of the production system." Laboratory quality assessment systems emphasize measurement of precision and accuracy, the first two of the five PARCC descriptors. Precision is a measure of the repeatability of a laboratory process; accuracy measures the closeness of an analytical result to the true or expected value.

Statistical Analysis

The heart of laboratory quality assessment is statistical analysis. To proceed with statistical analysis, a statistical model for the data must be chosen. Large sets of laboratory data have been found to follow a normal (*Gaussian*) distribution. Successful use of a statistical model requires awareness of how well the model applies to the particular data set under consideration. Very small data sets, for example, may depart from a normal distribution. In such situations, it may be necessary to average or normalize data from small sets to apply the Gaussian model. An evaluation of the precision (repeatability) of an analytical measurement is based on repeated analysis of stable samples of known or unknown composition. Evaluating accuracy requires the repetitive analysis of samples of known composition. Different types of quality control (QC) samples are analyzed in the laboratory to develop a data set that can be analyzed statistically. The statistical analysis will yield measures of precision and accuracy.

Quality Control Samples

As samples are received in the production analytical laboratory, they are grouped into "batches" for analysis. A batch may contain one or more samples for a variety of reasons (desired turnaround time, sample holding time, or laboratory backlog, to name a few). Usually the method SOP limits the maximum number of samples that may be included in a batch. Depending on the particular analytical method and the project requirements, one or more "batch QC samples" are scheduled for analysis with each analytical batch. Table 10–2 summarizes the typical batch QC samples that may be included in the analysis of field samples.

Control Charts

Two types of control charts are used to present and evaluate the results of the analysis of batch QC samples. The first type is called a *property chart,* or an *X chart.* The X chart requires results from a sample of known composition and is used to eval-

TABLE 10–2
Typical batch QC samples.

Batch QC Sample	Description	Purpose	Measures
Method or preparation blank	Contains no sample material but is treated like a sample in every other way.	Monitors any contamination to which the analytical batch may have been exposed during analysis. An acceptable result is below the level specified in the QAPP or the QAPjP.	*Contamination*
Laboratory control sample (LCS)	A sample that contains the analyte(s) of interest in known amounts.	Used to monitor the success of the analysis in recovering the analyte(s) of interest from a familiar sample matrix. $\text{Recovery} = \frac{\text{Found value}}{\text{Expected value}} \times 100$	*Accuracy* of the laboratory's performance of the method
Method spike	A method blank that has had a known amount of the analyte(s) of interest added to it.	Used to monitor method accuracy when an LCS in not available.	*Accuracy* of the laboratory's performance of the method
Method duplicate (MD)	A second portion of a field sample taken for analysis.	When the result is compared to the original result from the filed sample, a measure of precision can be developed. The most common measure is relative percent difference (RPD). $\text{RPD} = \frac{\lvert\text{Field sample result} - \text{MD result}\rvert}{(\text{Average of field sample and MD results})} \times 100\%$	*Precision*
Matrix spike (MS)	A second portion of a field sample to which a known amount of the analyte(s) of interest has been added.	Used to monitor the ability of the analytical method to recover the analyte(s) of interest from that particular field sample matrix.	*Accuracy* of the laboratory's performance with that type of sample
Matrix spike duplicate (MSD)	A *third* portion of the same field sample to which the same amount of the analytes of interest has been added.	When the result is compared to the result from the MS, a measure of precision can be developed. MS/MSD results usually are reported as relative percent of difference. (See MD.)	*Precision*
Surrogates	Compounds similar to the analyte(s) of interest but that are known *not* to be present in the environment.	When available, are added to each field sample. Surrogate recovery is a measure of the accuracy of the method for that particular sample.	*Accuracy*

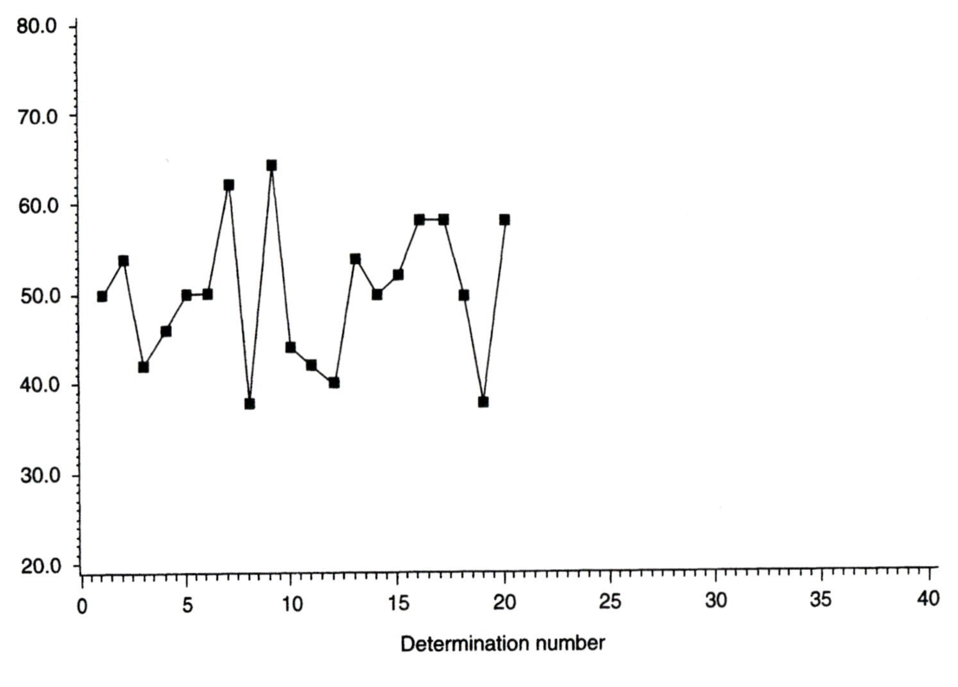

FIGURE 10–22
Run chart.

uate accuracy. The second type of chart is called the *precision chart*, or the *R chart*. It displays the range (typically the relative percent difference) of two or more measurements and is used to evaluate precision.

The simplest form of a property chart is the *run chart*. The run chart simply records the property of interest (recovery of an analyte) in a running sequence. Figure 10–22 illustrates a run chart. If the results in the series are averaged and the average value plotted as a center line, the property chart becomes more useful for evaluating results.

Figure 10–23 illustrates the run chart with a center line included. If, in addition to the average, the standard deviation of the results in the sequence is calculated, an X chart with warning and control limits can be constructed. Figure 10–24 illustrates an X chart. The warning limits are set at ±2 standard deviations. The control limits are set at ±3 standard deviations. If the values in series of results are in statistical control (that is, they fit the Gaussian statistical model—are normally distributed), 95 percent of all the values should fall between the upper warning limit and the lower warning limit. Additionally, 99.7 percent of the values should be found between the upper control limit and the lower control limit.

When an analytical method goes "out of control" (no longer fits the statistical model), it is likely that the analytical method has a problem that must be investigated and corrected. Such a situation is indicated by a change in the pattern of the X-type control chart. Figure 10–25 illustrates four common out-of-control patterns. Figure 10–26 illustrates three less common out-of-control patterns.

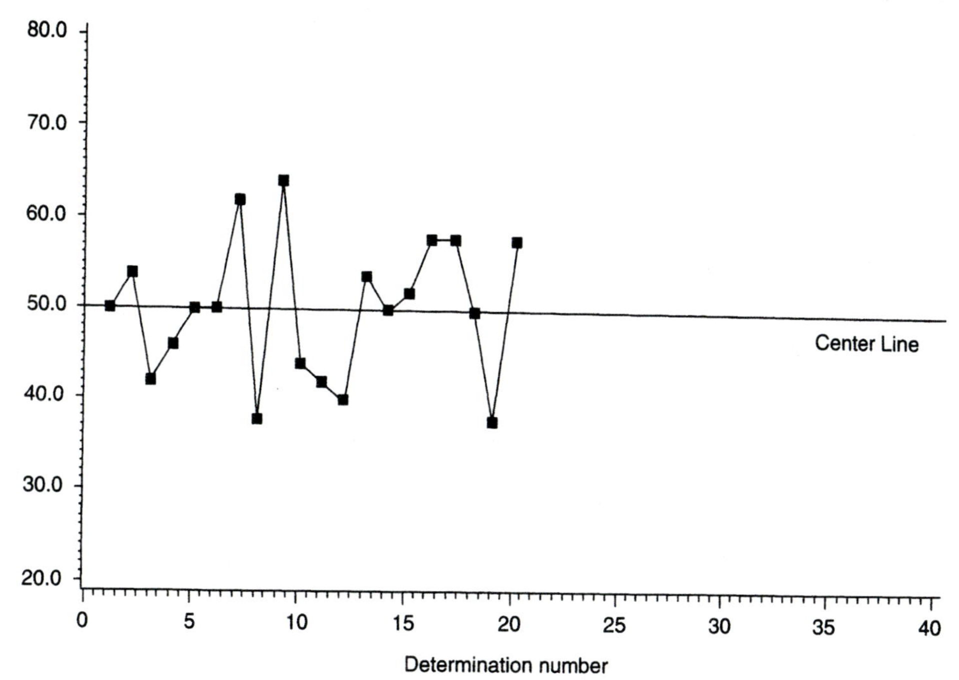

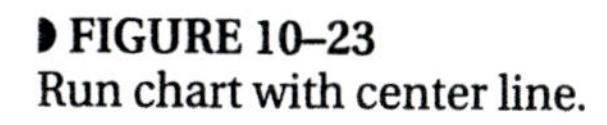

FIGURE 10–23
Run chart with center line.

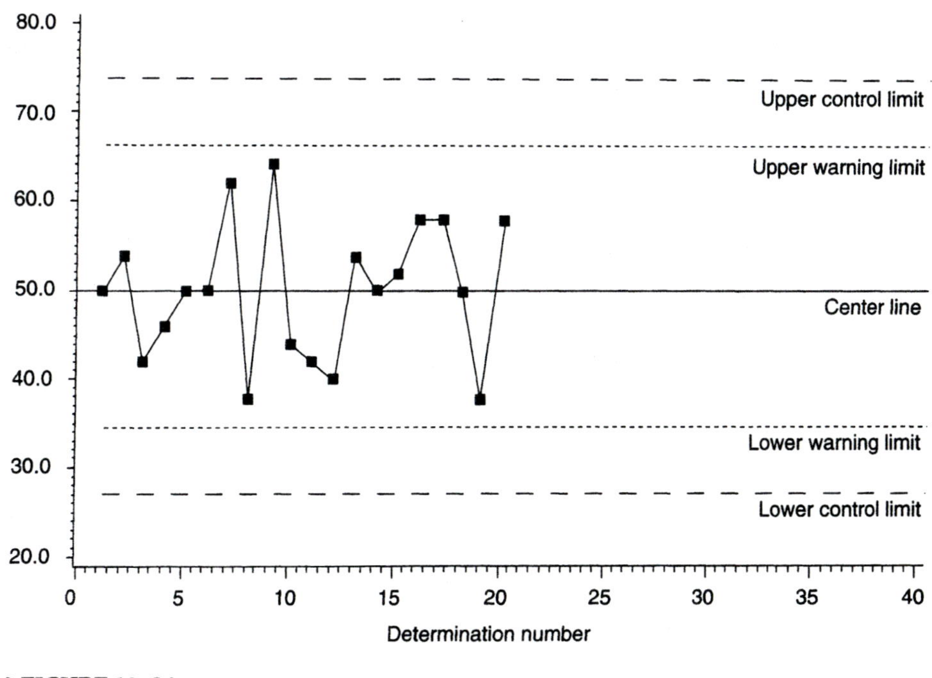

FIGURE 10–24
X chart.

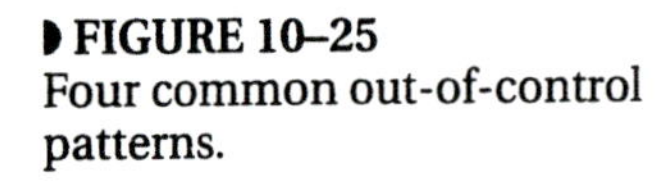

FIGURE 10–25
Four common out-of-control patterns.

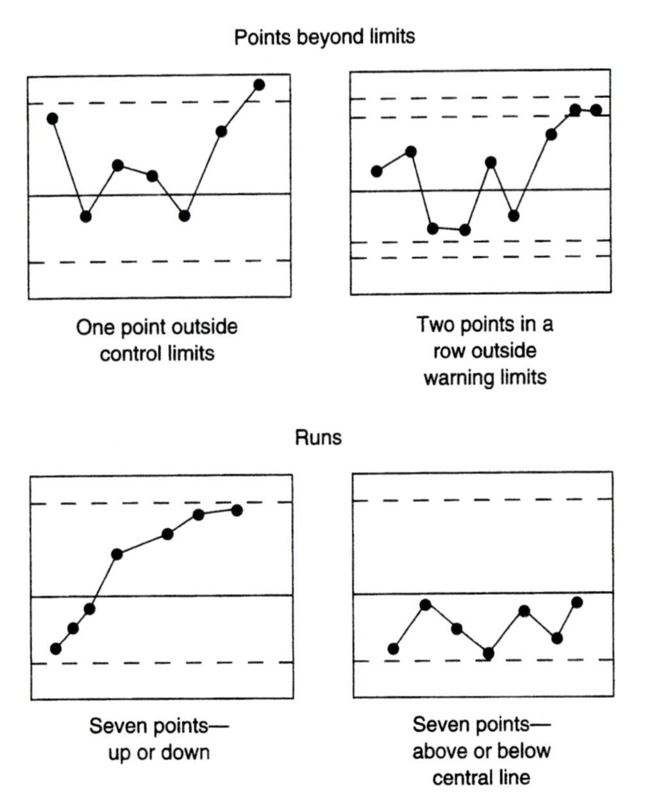

The precision- or range-type of control chart is used to monitor the repeatability of an analytical method. The relative percent difference is calculated and plotted as a sequence on the chart. The values are averaged to calculate a center line. For duplicate pairs of data, the value of the center line is multiplied by 2.512 to obtain the upper warning limit. The value of the center line is multiplied by 3.267 to obtain the upper control limit. The lower warning limit and the lower control limit are zero. Figure 10–27 illustrates an R-type chart for duplicate measurements. The R chart is used in a fashion similar to that for an X chart. If the data are in statistical control, 95 percent of all the values should be found between zero and the upper warning limit, and 99.7 percent of the points should lie between zero and the upper control limit.

The following situations observed on an R chart indicate that the precision of an analytical method is out of control:

- One point outside the upper control limit.
- Two successive points outside the upper warning limit.
- Five successive points trending in one direction.
- Seven successive points on one side of the center line.
- Cyclic patterns.
- Jumps or shifts.

FIGURE 10–26
Less common out-of-control patterns.

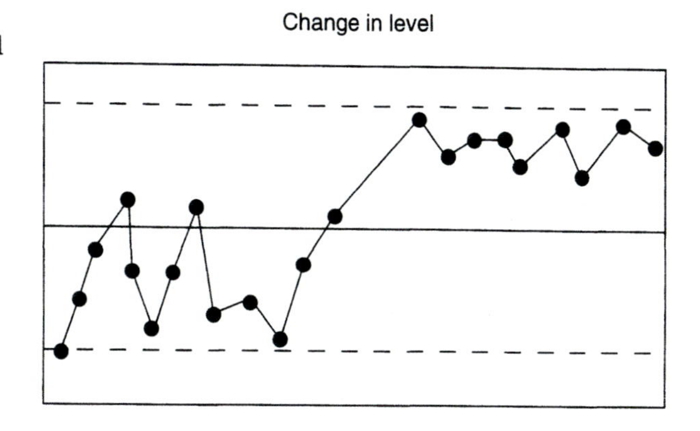

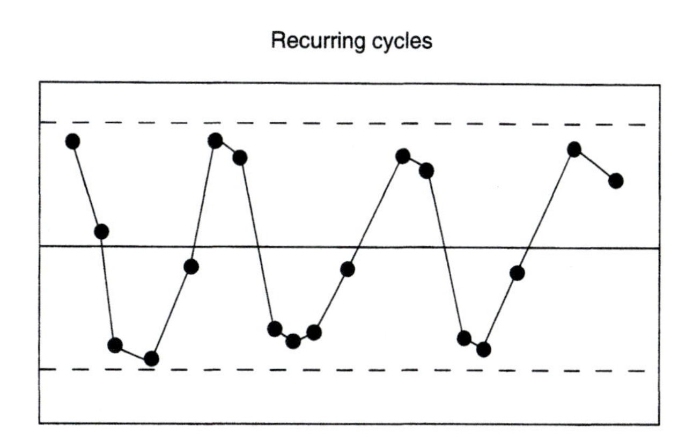

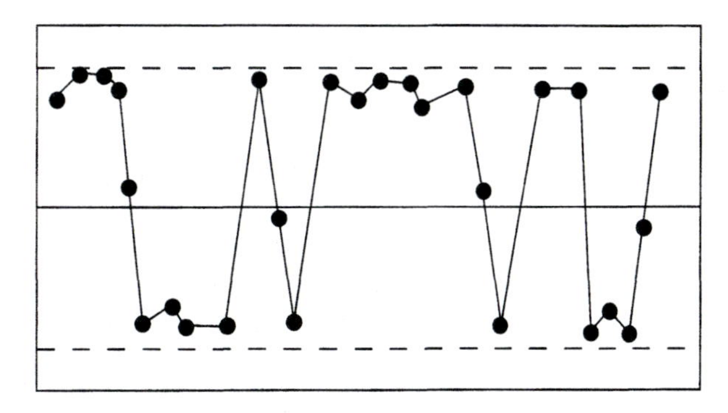

When the X or R control charts indicate that the accuracy or precision of an analytical method is out of control, the laboratory must suspend the use of the analytical method until the problem is identified and corrected. Most data users expect the laboratory to document the problem and its corrective action and then to report that information with the analytical results.

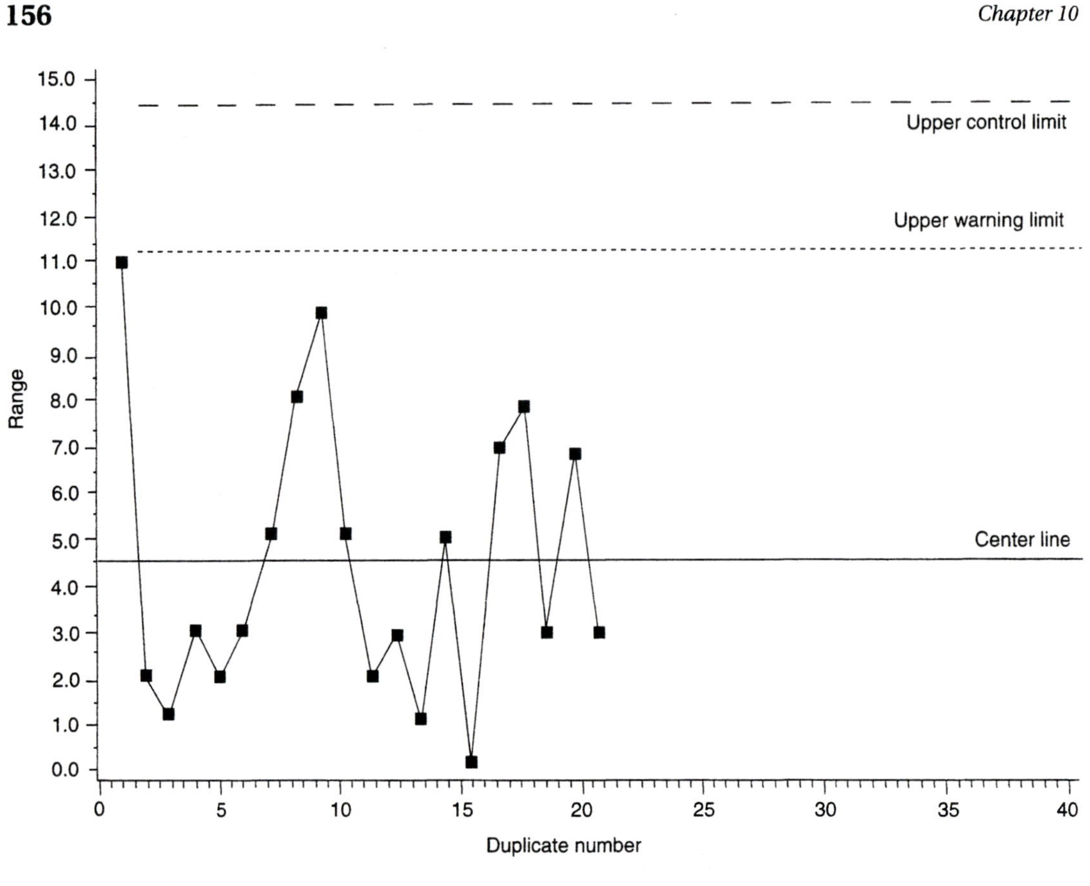

FIGURE 10–27
R chart for duplicate measurements.

EFFECTIVE COMMUNICATION WITH THE LABORATORY

As noted earlier, the Environmental Monitoring Methods Index (EMMI) allows searches of fifty EPA regulatory lists, 2,600 substances, and 926 analytical methods. Depending on the requirements of the project, different QA/QC strategies are applied to the analytical methods. The resulting potential for complexity is truly remarkable. The importance of effective communication with the laboratory cannot be overemphasized. Ineffective communication results in the production of unusable data, missed project deadlines, and cost overruns. Selecting a good analytical laboratory is the first step toward obtaining usable analytical results. Instructing an unknown laboratory to follow a good method does not guarantee good results. Only good laboratories produce good results. It is best to visit a laboratory before sending samples to it. Look for these things on the visit:

- Does the laboratory have an up-to-date QAPP? Will the laboratory give you a copy to review?
- Is the laboratory up-to-date with its accreditation and certifications?
- How has the laboratory scored on recent external performance evaluation samples?

- Does the laboratory have adequate staff and equipment in place?
- Does the laboratory offer you a single point of contact?

It is wise to ask other users of the laboratory's services about its reputation for quality and meeting turnaround-time requirements. Low price alone is not an adequate criterion for selecting a laboratory.

After selecting an analytical laboratory, the next step is to begin communication early in the project, preferably during the planning phase. Staff members of a good laboratory are able and willing to review analytical plans and make suggestions to improve quality and cost efficiency. Early communication also allows the laboratory to plan its resources so that it can meet turnaround times.

Communication with the laboratory also must be clear. It is entirely appropriate to give the laboratory a copy of the sampling and analysis plan. The sampling and analysis plan should address at least the following points:

- Reason for the project (the decision to be made).
- Goals for data quality.
- Sampling strategy (sampling points and number of samples to be collected).
- Sampling procedures.
- Sample storage and transportation.
- Laboratory analyses.
- Analytical reports.

For the laboratory in particular, the sampling and analysis plan should clearly specify the following items:

- The exact identification of each analytical method to be used.
- A clear list of the compounds to be determined by each analytical method, along with their detection limits.
- Number and type (water, soil, etc.) of samples to be analyzed by each method.
- When the samples will arrive at the laboratory (will they come at night or on the weekend?).
- The time allowed in the schedule for the laboratory to report results.
- The format and content of the analytical report.
- A list of such things as sampling containers and preservatives that the laboratory is expected to provide.

In addition to clear, early communication with a good laboratory, clear, ongoing communication is essential for the project's successful completion. Each group of samples sent to the laboratory for analysis should be accompanied by a written form that clearly identifies each sample with a unique number and specifies the analyses to be performed for each. Currently, the trend is to use chain-of-custody forms to accomplish this type of communication.

SUMMARY

Laboratory analysis is one step in a sequence of processes performed to provide information that will be used to make a decision regarding the environment. A large variety of regulatory programs, analytical methods, and regulated compounds, as well as different options for QA/QC, create a complex variety of options for laboratory analysis. A number of standard analytical methods (protocols) have been developed. Particular regulatory programs usually draw from a particular group of standard analytical methods. For example, the *SW-846* method manual contains the analytical procedures generally used to make decisions under RCRA regulations.

It is important to clearly specify to the analytical laboratory which analytical method is desired and which QA/QC strategy is expected.

The following items summarize key ideas in this chapter:

- Laboratory analysis provides information for the environmental decision process.
- A variety of environmental laws regulates 2,600 hazardous substances.
- Hundreds of analytical methods and a variety of QA/QC options are available.
- Each project requires selected analytical methods and QA/QC measures.
- Laboratories will provide useful analytical results when the methods and the QA/QC approach are specified clearly before the work starts.

QUESTIONS FOR REVIEW

1. How do the results from laboratory analyses contribute to environment protection?
2. Why are manuals of standardized laboratory methods (protocols) important?
3. How does a production analytical laboratory differ from a research analytical laboratory?
4. What minimum services can you expect from an environmental analytical laboratory?
5. What basic information does the laboratory need to complete sample analysis satisfactorily?
6. What is a flame atomic absorption (FAA) spectrophotometer? Draw a schematic of one and label the components. Why is the monochromator important?
7. What purpose does acid digestion serve when analyzing samples for metals?
8. What is a gas chromatograph? Draw a schematic of one and label the components. What is the role of each component in producing a chromatogram?
9. What types of information can be gained from a chromatogram?
10. What advantage does gas chromatography/mass spectrometry (GC/MS) have over other gas chromatographic methods?
11. What is the difference between quality control (QC) and quality assurance (QA) in the analytical laboratory?
12. What is the purpose of each of the following QC samples:
    - Method blank?
    - Laboratory control sample?
    - Matrix spike?
    - Matrix spike duplicate?

ACTIVITIES

1. Visit an environmental analytical laboratory and ask to follow the path a sample would take through the laboratory. What methods does the laboratory perform? What approach does the laboratory take to quality assurance? What types of QC samples does the laboratory analyze? How frequently (i.e., 1 in 10, 1 in 20) are QC samples analyzed?
2. Visit a consulting engineering firm that performs contamination assessments or environmental cleanups. Ask how analytical requirements are communicated to the laboratory. Are written quality assurance project plans available for the projects? Do you think that the requirements are stated clearly enough by the engineering firm for the laboratory to perform satisfactorily?

READINGS/RESOURCES

Standard Methods for the Examination of Water and Wastewater, 19th ed.

Methods for the Chemical Analysis of Water and Wastes, 1979. EPA-600/4-79-020. (Also refer to updates released in 1983, 1991, and 1993.)

Test Methods for Evaluating Solid Waste (SW-846), 3rd ed., 1986. EPA Office of Solid Waste. (Also refer to updates I and II, released in 1992 and 1995.)

Taylor, J. K., 1987. *Quality Assurance of Chemical Measurements*, Lewis Publishers.

APPENDIX 10–1: LIST OF REFERENCED ANALYTICAL METHODS*

Method 325.2—Chloride, colorimetric, automated ferricyanide, AAII.
Method 335.3—Cyanide, total, colorimetric, automated UV.
Method 350.1—Ammonia, colorimetric, automated phenate.
Method 353.2—Nitrate-nitrite, colorimetric, automated cadmium reduction.
Method 365.4—Phosphorus, total, colorimetric, automated, block digestor, AAII.
Method 370.1—Silica, dissolved, colorimetric.
Method 375.2—Sulfate, colorimetric, automated methyl thymol blue, AAII.
Method 150.1—pH, electrometric.
Method 340.2—Fluoride, potentiometric, ion-selective electrode.
Method 160.3—Residue, total, gravimetric, dried at 103–105°C.
Method 413.1—Oil and grease, total recoverable, gravimetric, separatory funnel extraction.
Method 376.1—Sulfide, titrimetric, iodine.
Method 377.1—Sulfite, titrimetric.
Method 6010A—Inductively coupled plasma—atomic emission spectroscopy.
Method 7060—Arsenic (atomic absorption, furnace technique).
Method 7421—Lead (atomic absorption, furnace technique).
Method 7740—Selenium (atomic absorption, furnace technique).
Method 7470A—Mercury in liquid waste (manual cold-vapor technique).
Method 7471—Mercury in solid or semisolid waste (manual cold-vapor technique).
Method 8070—Nitrosamines by gas chromatography.
Method 8141A—Organophosphorus compounds by gas chromatography: capillary column technique.
Method 8080A—Organochlorine pesticides and polychlorinated biphenyls by gas chromatography.
Method 8270B—Semivolatile organic compounds by gas chromatography/mass spectrometry (GC/MS): capillary column technique.
Method 8260A—Volatile organic compounds by gas chromatography/mass spectrometry (GC/MS): capillary technique.
Method 300.0—The determination of inorganic anions in water by ion chromatography.
Method 8310—Polynuclear aromatic hydrocarbons.
Method 8330—Nitroaromatics and nitramines by high-performance liquid chromatography (HPLC).

*Analytical methods are listed in the order in which they appear in the chapter.

APPENDIX 10–2: METHOD SUMMARIES

Method	325.2—Chloride, Colorimetric, Automated Ferricyanide, AAII
Reference	*Methods for Chemical Analysis of Water and Wastes* (EPA-600/4-79-020, revised March 1983)

Legislation SDWA and CWA

Sample matrix Water and wastewater

Description An aliquot of 50 to 100 mL is filtered and then used to rinse and fill the autosampler cup(s). Thiocyanate ion (SCN) is liberated from mercuric thiocyanate through sequestration of mercury by chloride ion to form un-ionized mercuric chloride. In the presence of ferric ion, the liberated SCN forms highly colored ferric thiocyanate in concentration proportional to the original chloride concentration.

Performance 1 to 200 mg chloride per liter

Sample container 125 mL plastic or glass bottle

Preservation No special requirements

Holding Time 28 days

Method 335.3—Cyanide, Total, Colorimetric, Automated UV

Reference *Methods for Chemical Analysis of Water and Wastes* (EPA-600/4-79-020, revised March 1983)

Legislation SDWA and CWA

Sample matrix Water and wastewater

Description An aliquot of 50 to 100 mL is filtered and then used to rinse and fill the autosampler cup(s). The cyanide as hydrocyanic acid (HCN) is released from cyanide complexes by means of UV digestion and distillation. Cyanides are converted to cyanogen chloride by reactions with chloramine-T, which subsequently reacts with pyridine and barbituric acid to give a red-colored complex.

Performance 5 to 500 micrograms cyanide per liter

Sample container 1 liter plastic or glass bottle

Preservation Add sodium hydroxide to bring pH above 12 and store at 4°C. If the sample tests positive in the field for oxidizers such as chlorine, add ascorbic acid crystals.

Holding time 14 days

Method 350.1—Ammonia, Colorimetric, Automated Phenate

Reference *Methods for Chemical Analysis of Water and Wastes* (EPA-600/4-79-020, revised March 1983)

Legislation SDWA and CWA

Sample matrix Water and wastewater

Description An aliquot of 50 to 100 mL is filtered and then used to rinse and fill the autosampler cup(s). Alkaline phenol and hypochlorite react with ammonia to form indophenol blue that is proportional to the ammonia concentration. The blue color formed is intensified with sodium nitroprusside.

Performance 0.01 to 2.0 mg per liter ammonia reported as nitrogen

Sample container 125 mL plastic or glass bottle

Preservation Adjust pH to less than 2 with sulfuric acid, store at 4°C

Holding time 28 days. (Note: Heavily polluted samples may require shorter holding times)

Method 353.2—Nitrate-Nitrite, Colorimetric, Automated Cadmium Reduction

Reference *Methods for Chemical Analysis of Water and Wastes* (EPA-600/4-79-020, revised March 1983)

Legislation	SDWA and CWA
Sample matrix	Water and wastewater
Description	A 50- to 100-mL portion is filtered and then used to rinse and fill the autosampler cup(s). Subsequently, each sample is passed through a column containing granulated copper-cadmium (Cu-Cd) to reduce nitrate to nitrite. The nitrite (that originally present plus reduced nitrate) is determined by diazotizing with sulfanilamide and coupling with N—(1-naphthyl)—ethylenediamine dihidrochloride to form a highly colored azo dye that is measured colorimetrically. Separate, rather than combined, nitrate-nitrite values are obtained readily by carrying out the procedure first with, and then without, the Cu-Cd reduction step.
Performance	0.05 to 10.0 mg. per liter nitrate plus nitrite, reported as nitrogen
Sample container	125 mL plastic or glass bottle
Preservation	Reduce pH to less than 2 with sulfuric acid, store at 4°C
Holding time	28 days

Method	365.4—Phosphorus, Total, Colorimetric, Automated, Block Digestor, AAII
Reference	*Methods for Chemical Analysis of Water and Wastes* (EPA-600/4-79-020, revised March 1983)
Legislation	SDWA and CWA
Sample matrix	Water and wastewater
Description	A 25-mL portion of the sample is digested with a solution of sulfuric acid, mercuric sulfate, and potassium sulfate. The digestion is performed for 1 hour at 160°C and then for $2^1/2$ hours at 380°C. The residue is cooled, diluted, and analyzed by Auto Analyzer using a molybdate color reagent.
Performance	0.01 to 20 mg. phosphate per liter, reported as phosphorus
Sample container	125 mL plastic or glass bottle
Preservation	Reduce pH to less than 2 with sulfuric acid, store at 4°C
Holding time	28 days

Method	370.1—Silica, Dissolved, Colorimetric
Reference	*Methods for Chemical Analysis of Water and Wastes* (EPA-600/4-79-020, revised March 1983)
Legislation	SDWA and CWA
Sample matrix	Water and wastewater
Description	The sample is filtered and a 50-mL portion is taken for analysis. After digestion the high-range color standard, molybdate ion is added. Samples that show color in the high range are measured using a spectrophotometer. The low-range color reagent (1-amino-2-naphthol-4-sulfonic acid) may be added to samples that do not have observable color in the high range. When the blue, low-range color develops, it is measured in the spectrophotometer at a different wavelength.
Performance	High range—2 to 25 mg silica per liter Low range:—0.1 to 2 mg silica per liter
Sample container	250 mL plastic bottle
Preservation	No special requirements
Holding time	28 days

Method	375.2—Sulfate, Colorimetric, Automated Methylthymol Blue, AAII
Reference	*Methods for Chemical Analysis of Water and Wastes* (EPA-600/4-79-020, revised March 1983)
Legislation	SDWA and CWA
Sample matrix	Water and wastewater
Description	The sample is first passed through a sodium form cation-exchange column to remove multivalent metal ions. The sample containing sulfate is then reacted with an alcohol solution of barium chloride and methylthymol blue (MTB) at a pH of 2.5–3.0 to form barium sulfate. The combined solution is raised to a pH of 12.5–13.0 so that excess barium reacts with MTB. The uncomplexed MTB color is gray; if it is all chelated with barium, the color is blue. Initially, the barium and MTB are equimolar and equivalent to 300 mg SO_4/1; thus, the amount of uncomplexed MTB is equal to the sulfate present.
Performance	3 to 300 mg sulfate per liter
Sample container	125 mL plastic or glass bottle
Preservation	No special requirements
Holding time	28 days

Method	150.1—pH, Electrometric
Reference	*Methods for Chemical Analysis of Water and Wastes* (EPA-600/4-79-020, revised March 1983)
Legislation	SDWA and CWA
Sample matrix	Water and wastewater
Description	The pH of a sample is determined electrometrically using either a glass electrode in combination with a reference potential electrode or a single-probe combination-type electrode.
Performance	pH is reported to the nearest 0.1 unit.
Sample container	125 mL plastic bottle
Preservation	Samples for pH should be analyzed as soon as possible, preferably in the field at the time of sampling. If the samples must be taken to the laboratory for analysis, the sample containers should be filled completely and kept sealed prior to analysis. The pH of a water sample not at equilibrium with the atmosphere can change when exposed to the atmosphere.
Holding time	Analyze immediately

Method	340.2—Fluoride, Potentiometric, Ion-Selective Electrode
Reference	*Methods for Chemical Analysis of Water and Wastes* (EPA-600/4-79-020, revised March 1983)
Legislation	SDWA and CWA
Sample matrix	Water and wastewater
Description	Fluoride is determined potentiometrically using a fluoride electrode and a selective ion meter having a direct concentration scale for fluoride. Fluoride also can be determined using the fluoride electrode in conjunction with a standard reference electrode and a pH meter having an expanded millivolt scale.
Performance	0.1 to 1,000 mg per liter fluoride

Sample container 125 mL plastic bottle
Preservation No special requirements
Holding time 28 days

Method 160.3—Residue, Total, Gravimetric, Dried at 103–105°C
Reference *Methods for Chemical Analysis of Water and Wastes* (EPA-600/4-79-020, revised March 1983)
Legislation SDWA and CWA
Sample matrix Water and wastewater
Description A well-mixed portion of the sample is quantitatively transferred to a preweighed evaporating dish and evaporated to dryness at 103–105°C. The size of the portion is chosen to ensure that at least 25 mg of residue are present.
Performance 10 to 20,000 mg per liter total residue
Sample container 250 mL plastic or glass bottle
Preservation Cool to 4°C
Holding time 7 days

Method 413.1—Oil and Grease, Total Recoverable, Gravimetric, Separatory Funnel Extraction
Reference *Methods for Chemical Analysis of Water and Wastes* (EPA-600/4-79-020, revised March 1983)
Legislation SDWA and CWA
Sample matrix Water and wastewater
Description The sample is acidified to a pH less than 2 and serially extracted with fluorocarbon-113 in a separatory funnel. The solvent is evaporated from the extract and the residue is weighed. To reduce the amount of fluorocarbon-113 released to the atmosphere, this method will be modified in the near future by the EPA to use hexane as the extraction solvent.
Performance 5 to 1,000 mg extractable material per liter
Sample container 1 liter glass bottle
Preservation Adjust pH to less than 2 with sulfuric or hydrochloric acid, store at 4°C
Holding time 28 days

Method 376.1—Sulfide, Titrimetric, Iodine
Reference *Methods for Chemical Analysis of Water and Wastes* (EPA-600/4-79-020, revised March 1983)
Legislation SDWA and CWA
Sample matrix Water and wastewater
Description Sulfide can be lost from the sample as hydrogen sulfide. Zinc acetate is added at sampling to preserve the sulfude as zinc sulfide. Excess iodine is added to the sample to react with the zinc sulfide formed during sampling. The iodine oxidizes the sulfide to sulfur under acidic conditions. The excess iodine is backtitrated with sodium thiosulfate or phenylarsine oxide.
Performance This method is suitable for the measurement of sulfide in concentrations above 1 mg per liter.

Sample container 1 liter plastic or glass bottle
Preservation Adjust pH to greater than 9 using sodium hydroxide; add zinc acetate and store at 4°C
Holding time 7 days

Method 377.1—Sulfite, Titrimetric
Reference *Methods for Chemical Analysis of Water and Wastes* (EPA-600/4-79-020, revised March 1983)
Legislation SDWA and CWA
Sample matrix Water and wastewater
Description A 50-mL portion of sample is acidified and titrated with a standard potassium iodide-iodate titrant. The faint permanent blue endpoint appears when the reducing power of the sample has been completely exhausted.
Performance This method is suitable for the measurement of sulfite in concentrations greater than 2 mg sulfite per liter.
Sample container 250 mL plastic or glass bottle
Preservation No special requirements
Holding time Analyze immediately

Method 6010A—Inductively Coupled Plasma—Atomic Emission Spectroscopy
Reference *Test Methods for Evaluating Solid Waste (SW-846)*
Legislation RCRA
Sample matrix Water, wastes, soils, sludges, and sediments
Description Prior to analysis, the sample must be solubilized or digested using an appropriate sample-preparation method. The purpose of the digestion step is to dissolve the metals in the sample in strong acids. Method 6010 describes the multielemental determination of metals by inductively coupled plasma (ICP). The method measures the light emitted by the different elements by means of optical spectrometry. The wavelength of the emitted light is used to identify the metal, and the intensity of the emitted light is used to measure the amount of the metal present.
Performance The method detection limit ranges from 0.3 micrograms per liter in water for beryllium up to 75 micrograms per liter for selenium. The upper end of the range is 10,000 to 20,000 micrograms per liter.
Sample container 500 mL plastic or glass bottle. Plastic typically is used.
Preservation Reduce pH to less than 2 with nitric acid
Holding time 6 months

Method 7060—Arsenic (Atomic Absorption, Furnace Technique)
Reference *Test Methods for Evaluating Solid Waste (SW-846)*
Legislation RCRA
Sample matrix Wastes, soils, and water
Description Prior to analysis, the samples must be digested in a mixture of nitric acid and hydrogen peroxide. The purpose of this digestion is to convert organic forms of arsenic to inorganic forms and to dissolve the arsenic in the acid solution for

	analysis. The analysis is carried out using graphite furnace atomic absorption spectroscopy.
Performance	A lower limit of 1 to 5 micrograms arsenic per liter in water is typical. The upper range is 100 micrograms arsenic per liter.
Sample container	500 mL plastic or glass bottle. Plastic typically is used.
Preservation	Reduce pH to less than 2 with nitric acid
Holding time	6 months

Method	7421—Lead (Atomic Absorption, Furnace Technique)
Reference	*Test Methods for Evaluating Solid Waste (SW-846)*
Legislation	RCRA
Sample matrix	Wastes, soils, and water
Description	Prior to analysis, the samples must be digested in a mixture of nitric acid and hydrogen peroxide. The purpose of this digestion is to convert organic forms of lead to inorganic forms and to dissolve the lead in the acid solution for analysis. The analysis is carried out using graphite furnace atomic absorption spectroscopy.
Performance	A lower limit of 1 to 5 micrograms lead per liter in water is typical. The upper range is 100 micrograms lead per liter.
Sample container	500 mL plastic or glass bottle. Plastic typically is used.
Preservation	Reduce pH to less than 2 with nitric acid
Holding time	6 months

Method	7740—Selenium (Atomic Absorption, Furnace Technique)
Reference	*Test Methods for Evaluating Solid Waste (SW-846)*
Legislation	RCRA
Sample matrix	Wastes, soils, and water
Description	Prior to analysis, the samples must be digested in a mixture of nitric acid and hydrogen peroxide. The purpose of this digestion is to convert organic forms of selenium to inorganic forms and to dissolve the selenium in the acid solution for analysis. The analysis is carried out using graphite furnace atomic absorption spectroscopy.
Performance	A lower limit of 1 to 5 micrograms selenium per liter in water is typical. The upper range is 100 micrograms selenium per liter.
Sample container	500 mL plastic or glass bottle. Plastic typically is used.
Preservation	Reduce pH to less than 2 with nitric acid
Holding time	6 months

Method	7470A—Mercury in Liquid Waste (Manual Cold-Vapor Technique)
Reference	*Test Methods for Evaluating Solid Waste (SW-846)*
Legislation	RCRA
Sample matrix	Groundwater and aqueous wastes
Description	A 100-milliliter portion of the sample is digested with sulfuric acid, nitric acid, and potassium permanganate. The dissolved mercury ions are reduced to elemental mercury and measured by means of cold-vapor atomic absorption spectroscopy.
Performance	The typical detection limit for mercury is 0.2 micrograms mercury per liter. The upper limit is 10 micrograms mercury per liter.

Sample container 500 mL glass or plastic bottle
Preservation Adjust pH to less than 2 with nitric acid
Holding time 28 days

Method 7471—Mercury in Solid or Semisolid Waste (Manual Cold-Vapor Technique)
Reference *Test Methods for Evaluating Solid Waste (SW-846)*
Legislation RCRA
Sample matrix Soils, sediments, and sludge-type materials
Description Triplicate 0.2-gram portions of the sample are weighed and placed in digestion vessels. The samples are digested with aqua regia and potassium permanganate. The dissolved mercury is reduced to elemental mercury and analyzed by cold-vapor atomic absorption spectroscopy.
Performance The range for this method is 0.1 to 5 micrograms mercury per gram.
Sample container Glass or plastic widemouthed jar
Preservation Store at 4°C
Holding time 28 days

Method 8070—Nitrosamines by Gas Chromatography
Reference *Test Methods for Evaluating Solid Waste (SW-846)*
Legislation RCRA
Sample matrix Water and wastewater
Description One liter of aqueous sample is solvent-extracted with methylene chloride using a separatory funnel. The methylene chloride extract is washed with dilute HCl to remove free amines, dried, and concentrated to a volume of 10 mL or less. The extract is analyzed by gas chromatography using a nitrogen phosphorus detector (GC/NPD).
Performance Detection limits for the three nitrosamines are near 1 microgram per liter.
Sample container 1 liter glass bottle with Teflon®-lined cap or glass soil jar with Teflon-lined cap
Preservation Store at 4°C
Holding time 7 days between sample collection and extraction; 40 days between extraction and analysis

Method 8141A—Organophosphorus Compounds by Gas Chromatography: Capillary Column Technique
Reference *Test Methods for Evaluating Solid Waste (SW-846)*
Legislation RCRA
Sample matrix Water and soil
Description This method has been validated for 26 phosphorus-containing pesticides and herbicides. It has been applied to a number of other related compounds. Aqueous samples are extracted at a neutral pH with methylene chloride. Solid samples are prepared for analysis by means of a Soxhlet extraction with methylene chloride/acetone (1:1). A gas chromatograph equipped with a flame photometric detector (GC/FPD) is used to complete the analysis.

Performance	Water detection limits range between 0.04 and 0.8 micrograms per liter. Soil detection limits range between 2 and 40 micrograms per kilogram. The upper range for water samples is 50 micrograms per liter, and the upper range for soil is 1700 micrograms per kilogram.
Sample container	1 liter glass bottle with Teflon®-lined cap or glass soil jar with Teflon-lined cap.
Preservation	Store at 4°C
Holding time	7 days between sample collection and extraction; 40 days between extraction and analysis

Method	8080A—Organochlorine Pesticides and PCBs
Reference	*Test Methods for Evaluating Solid Waste (SW-846)*
Legislation	RCRA
Sample matrix	Water
Description	Up to 26 chlorine-containing pesticides and PCBs can be determined by this Method. Some of the analytes are aldrin, lindane, DDT, dieldrin, heptachlor, toxaphene, and aroclor-1260 (one of the PCBs). One liter of water is extracted with methylene chloride. The extract is reduced in volume by evaporation. The methylene chloride solvent is exchanged for hexane and the final volume brought to 10.0 mL. The extracts are analyzed by gas chromatography with an electron capture detector (GC/ECD).
Performance	Method detection limits range from 0.002 μg/L for dieldrin to 0.24 μg/L for toxaphene. Recoveries can range from 8% to 215%, depending upon the compound.
Sample container	1-liter amber glass bottle with a Teflon®-lined cap
Preservation	Store at 4°C; add sodium thiosulfate if residual chlorine is present
Holding time	7 days from sample collection to extraction; 40 days from extraction to analysis

Method	8270B—Semivolatile Organic Compounds by Gas Chromatography/Mass Spectrometry (GC/MS): Capillary Column Technique
Reference	*Test Methods for Evaluating Solid Waste (SW-846)*
Legislation	RCRA
Sample matrix	Solid waste, soil, and water
Description	This method has been applied to approximately 200 different compounds. However, 120 compounds or less typically are requested for routine analysis. Water samples are prepared for analysis by extracting the acidic compounds and the basic/neutral compounds at different pHs with methylene chloride. The extracts are combined and reduced in volume. Solid samples are prepared for analysis by Soxhlet extraction or by sonication. The extracts are reduced in volume and analyzed by gas chromatography and mass spectrometry (GC/MS).
Performance	Detection limits range from 0.1 to 30 micrograms per liter in water and from 3 to 100 micrograms per kilogram in soil.
Sample container	1 liter glass with Teflon®-lined cap, or glass soil jar with Teflon-lined cap
Preservation	Store at 4°C
Holding time	7 days between sample collection and extraction; 40 days between extraction and analysis

Method 8260A—Volatile Organic Compounds by Gas Chromatography/Mass Spectrometry (GC/MS): Capillary Technique

Reference *Test Methods for Evaluating Solid Waste (SW-846)*

Legislation RCRA

Sample matrix Water, wastes, and soil

Description This method has been validated for 58 volatile organic compounds and has been applied to a number of additional compounds as well. The volatile compounds are introduced into the gas chromatograph by the purge-and-trap method (stripping out of the sample with helium, trapping on a sorbent, and subsequent thermal desorption onto the column). The compounds are separated by gas chromatography and introduced into a mass spectrometer for identification and quantitation.

Performance Detection limits for volatile compounds range from 0.5 to 2 micrograms per liter for volatile compounds in water.

Sample container Two 40 mL vials with Teflon®-lined septum caps

Preservation Adjust pH to less than 2 with HCl, $H2SO_4$, or $NaHSO_4$. Store at 4°C.

Holding time 14 days (7 days for samples without pH adjustment)

Method 300.0—The Determination of Inorganic Anions in Water by Ion Chromatography

Reference EPA-600/4-84-017, March 1984

Legislation SDWA

Sample matrix Drinking water (sometimes applied to groundwater and wastewater)

Description A small volume of sample is introduced into an ion chromatograph. The anions of interest are separated using ion exchange columns and measured using a conductivity detector.

Performance Detection limits for the ions range from 4 to 85 micrograms per liter in water.

Sample container 125 mL plastic bottle

Preservation Store at 4°C

Holding time 28 days for fluoride, chloride and sulfate; 48 hours for orthophosphate, nitrate, and nitrite

Method 8310—Polynuclear Aromatic Hydrocarbons

Reference *Test Methods for Evaluating Solid Waste (SW-846)*

Legislation RCRA

Sample matrix Water and wastes

Description Water samples are extracted at a neutral pH with methylene chloride. Solid samples are extracted by sonication or by the Soxhlet method. The extract volume is reduced and then exchanged to acetonitrile from methylene chloride. The acetonitrile extract is injected into a high-performance liquid chromatograph (HPLC) equipped with a fluorescence detector.

Performance Detection limits range from 0.02 to 1 microgram per liter in water.

Sample container 1 liter glass bottle with Teflon®-lined cap, or a soil jar with a Teflon-lined cap

Preservation Store at 4°C

Holding time 7 days between sample collection and extraction; 40 days between extraction and analysis

Method	8330—Nitroaromatics and Nitramines by high-performance liquid chromatography (HPLC)
Reference	*Test Methods for Evaluating Solid Waste (SW-846)*
Legislation	RCRA
Sample matrix	Water, soil, and sediment
Description	This method was designed to measure 14 compounds that include explosives and their degradation products. Water samples are extracted using acetonitrile and salting out with sodium chloride. Soil and sediment samples are extracted using acetonitrile in an ultrasonic bath. The acetonitrile extracts are analyzed by HPLC with an ultraviolet (UV) detector.
Performance	Detection limits range from 0.04 to 1 microgram per liter in water, and from 0.06 to 0.3 micrograms per gram in solids.
Sample container	1 liter glass bottle with Teflon®-lined cap or soil jar with Teflon-lined cap
Preservation	Store at 4°C
Holding time	7 days between sample collection and extraction; 40 days between extraction and analysis

11

Nondestructive Testing Technology

Mark Sabolik

Upon completion of this chapter, you will be able to do the following:

- Understand a variety of nondestructive testing techniques, including leak, ammonia and hydrochloric acid reaction, liquid penetrant, ultrasonic, eddy current, radiographic, and magnetic particle.
- Understand information that may be helpful in testing materials and components associated with hazardous materials and disposal.

INTRODUCTION TO NONDESTRUCTIVE TESTING

Nondestructive testing (NDT) can be defined as any test method performed on a part or article that does not impair its future usefulness. The usefulness of any nondestructive testing inspection relies on the ability of a specific test method to locate discontinuities or flaws. Of main importance are the size, shape, orientation, and location of these flaws and the way that they will affect the useful life cycle of the part.

The primary goal of an NDT technician is to detect flaws. Known as *in-process inspection*, this can be performed during any part of the manufacturing process. Testing the part after it has been processed is called *final inspection*. Testing a part after it has been placed in service is called *in-service inspection*.

A part may contain defects that are not rejectable according to the code or specification. Standards are established and used to calibrate the equipment and as a comparison to determine the severity of a particular flaw. This is based on the type and location of the flaw and the way it would affect the service life of the part.

To determine the proper test method to apply to detect flaws, one needs to have a clear understanding of each method and its advantages and disadvantages. In some cases, more than one test method is used to ensure complete and accurate inspection.

Nondestructive testing comprises many different testing methods. The most common are covered in this chapter. They include leak testing, radiography, liquid penetrant testing, ultrasonic testing, eddy current testing, and magnetic particle testing. Other test methods available but not discussed in this chapter are neutron radiography, thermography, acoustic emission testing, holography, acoustic microscopy, magabsorption, microwave inspection, and optical holography.

Some important terms used in NDT are as follows:

- *Flaw* A lack of continuity or an imperfection in a physical or dimensional attribute of the part.
- *Discontinuity* Any break or interruption in the normal physical structure of an article; therefore, a discontinuity may be acceptable or rejectable.
- *Relevant indications* Errors in the material that may or may not be considered a defect. These are evaluated by a close visual inspection and by studying the blueprints and all information that was obtained from the nondestructive testing method. More than one method may be used to clarify a relevant indication. All relevant indications are evaluated to the appropriate code, which determines the accept/reject criteria.
- *Nonrelevant indications* Indications that are not or cannot be associated with a discontinuity. Examples of nonrelevant indications include mill scale, part configuration, changes of component properties such as magnetic properties or conductivity, brazed joints, grain structure, and voids in press-fitted components.
- *Defect* Any type of discontinuity that may compromise the integrity of the part. Not all defects may be rejectable. This is determined by the applicable code.
- *Interpretation* The ability to determine the probable cause of the discontinuity and the type of defect it is. Evaluation involves the acceptance or rejection of the discontinuity. This is based on the accept/reject criteria of the applicable code assigned for a particular test or examination.

In most cases the procedures and acceptance specifications have been determined. There are a larger number of organizations that have established the acceptable procedures required for the inspection of parts. These include, but are not limited to, the following:

American Welding Society (AWS)
American National Standards Institute (ANSI)
American Society of Mechanical Engineers (ASME)
American Standards for Testing and Materials (ASTM)
American Petroleum Institute (API)
American Society for Metals (ASM)
Federal Aviation Administration (FAA)

In most cases the inspector does not write his or her own procedures and acceptance criteria, but follows the proper code and acceptance criteria.

Leak Testing

Leak testing is a nondestructive testing method that is used to detect and locate leaks. It is also used to measure the fluid leakage in pressurized or evacuates systems or components. A leak may be defined as any type of hole or flaw in a component that has allowed a liquid or gas to escape. A leak may be a crack, hole, fissure, or crevice. Any flaw in the final product may compromise the reliability and serviceability of the component. This test will verify that the part inspected is sound and that premature failure can be avoided. The three basic reasons that leak testing is performed are to:

1. Prevent any type of fluid or gas leakage that would interfere with the service reliability of a part.
2. Prevent any type of contamination of the parts and components adjacent to the part holding the fluid or gas.
3. Detect any part or components that would prematurely fail due to improper manufacturing or poor design.

The type and method of leak testing are determined by the level of sensitivity that is required. This level is determined by the engineers using the specifications outlined in the blueprints. The following are the three major factors used to determine the choice of testing:

1. The physical characteristics of the system and the tracer fluid used.
2. The type and size of the leak that would result in the component failing its intended purpose or function.
3. If a leak is located, what level is considered acceptable or rejectable, and whether the equipment allows for accurate, recordable measurements.

The basic terminology used to understand leak testing is very important. Strict definitions exist for the following:

- *Leak* An actual through-wall discontinuity or passage through which a fluid flows or permeates. A leak is simply a special type of flaw.
- *Leakage* The fluid that has flowed through a leak.
- *Leak rate* The amount of fluid passing through the leak per unit of time under a given set of conditions; properly expressed in units of quantity of mass per unit of time.
- *Minimum detectable leak* The smallest hole or discrete passage that can be detected.
- *Minimum detectable leak rate* The smallest detectable fluid-flow rate.

A leak is measured by how much leakage occurs under a specific set of conditions for a specific period of time. Leakage will vary depending on the conditions in which the test is performed; therefore, at a given temperature, the product of pressure and volume of a specific type of gas is proportional to its mass. This allows a measurement of pressure and volume per unit of time.

The following are the basic types of leaks:

Real leaks have a discrete passage through which fluid will flow. Examples are cracks or small ruptures in tubing. This type of leak may grow over time, depending on the conditions related to pressure, temperature, and cyclic or static loading.

Virtual leaks involve the gradual absorption of gases from surfaces or the escape of gases from a sealed component. Depending on temperature, pressure, and load, these type of leaks may become real leaks.

Leak testing may be as simple as flowing a soapy solution on the surface of a pressurized container and looking for bubbles produced from the escaping gases, or as detailed as measuring the gases using mass spectrometer instrumentation.

Bubble Leak Testing

The bubble method is a relatively simple nondestructive testing method used to detect leaks. It involves the immersion of a pressurized part into a tank of a liquid such as distilled or deionized water, treated waters, or hot oil. The part is submerged in the test liquid, and if there are any leaks, bubbles form at the exit points and rise to the surface of the immersion tank. Different levels of sensitivity can be obtained by submerging the part in different liquids because each type of liquid has a specific surface tension. Wetting agents can be added to the liquid to reduce the formation of bubbles on the surface of the part, which could be a source of false leak or false indication. Another way to increase the sensitivity is to submerge the part more deeply in the liquid.

Some vessels that require leak testing can exceed 20 feet in diameter. Vessels of this size are too large to submerge in a tank. In this situation, the vessel is pressurized from the inside, and a testing liquid is applied to the outside of it. Any leaking will create easily identifiable bubbles on the outside surface of the vessel. It is important that the technician flow the solution on the vessel; spraying the solution may cause bubbles to form during the application, leading to false indications.

The disadvantage of bubble leak testing is that it is not a highly sensitive test method. Therefore, it is used most often to detect gross discontinuities, because the size of the leak cannot be determined with accuracy. On the other hand, it is an inexpensive preliminary test that does not require highly skilled personnel. It also helps technicians repair or reject components early in the processing stage. Whenever you do use bubble testing, keep in mind the following limitations:

1. Contamination of the test surface will affect the accuracy of the test. Cleaning agents may clog or block potential leaks.
2. Very small or slow leaks may escape detection.
3. The viscosity of the test liquid may inhibit the ability to detect leakage.
4. The test liquids may become contaminated.
5. Out-gassing from corroded test surfaces may cause false indications.
6. Bubble leak tests may clog or block leakage and lower the sensitivity of an inspection using more sensitive testing methods later in the manufacturing process.

Ammonia and Hydrochloric Acid Reaction Testing

An *ammonia and hydrochloric acid reaction test* is a leak test method that requires good ventilation because of the hazards of hydrogen chloride vapor and ammonia gas. The test involves pressurizing a vessel with ammonia gas. After the vessel has reached the necessary pressure, the inspector will search for a leak with an open bottle of hydrochloric acid. If a leak is present, a white mist of ammonium chloride precipitate will appear.

False Readings and Errors Associated with Leak Testing

One of the most common mistakes associated with leak testing is using a test method that is too sensitive. If the code or procedure is not followed, an excessive number of components may be rejected. This becomes unnecessarily expensive. In some instances, more than one type of leak testing should be used. This is required if there are gross leaks that could mask smaller leaks in a vessel. In such cases, leak testing should be performed in two or more stages, beginning with gross leaks and ending with the most sensitive leak testing technique.

The last area that may lead to misinterpreting results involves the environment where the testing takes place. When tracer gases are used, the atmosphere should be monitored. Any type of contamination will show false or inaccurate readings. Background noise could be a potential problem when monitoring sonic or ultrasonic signatures. Airborne contaminants may cause temporary plugging of leaks of the vessel being inspected; this also could compromise an accurate leak test.

Liquid Penetrant Testing

Liquid penetrant testing was first performed during the days of locomotive steam engines. Large engine parts were immersed in kerosene for a period of time, allowing the kerosene to soak into any cracks. Kerosene was used because it was the only liquid available that had a good wetting ability and a low viscosity. Wetting ability prevented the part from drying, and the low viscosity allowed the kerosene to penetrate easily into any cracks. After the excess kerosene was removed, white chalk was dusted onto the surface of the part. The part was then struck with a hammer, which caused the part to vibrate and speed up the bleed out of oil that had soaked into any cracks. This method was acceptable for finding large cracks but had only limited success in locating small, fine ones.

Today, liquid penetrant testing has become one of the most common test methods performed. It involves the use of a visible or fluorescent dye that is sprayed, dipped, or wiped onto the part's surface. The penetrant is allowed to soak into any cracks or discontinuities open to the surface for a specific period called *dwell time*, which usually varies from 5 to 30 minutes. The excess penetrant is washed off with a coarse water spray, and the part is allowed to air dry or placed in a dryer.

The ability of a penetrant to perform properly is directly related to its *wetting ability*, which is the ability of a given material to flow over the surface of a part. This can be affected by the cleanliness of the part, the ability of the penetrant to wet the surface, the surface tension of the liquid, and the contact angle of the penetrant.

Wetting characteristics of penetrants are evaluated by the angle of the droplet in relationship to the surface of the part. A penetrant needs to enter small cracks and discontinuities and must have the ability to be retained in the flaws. After the appropriate dwell time, the penetrant should be able to soak back out of the flaws so they can easily be seen and evaluated. Figure 11–1 shows the basic steps used in penetrant testing.

FIGURE 11–1
Diagrams of the basic steps in penetrant testing:
(a) application of penetrant to the part,(b) removal of excess penetrant, and
(c) bleed out from the crack.

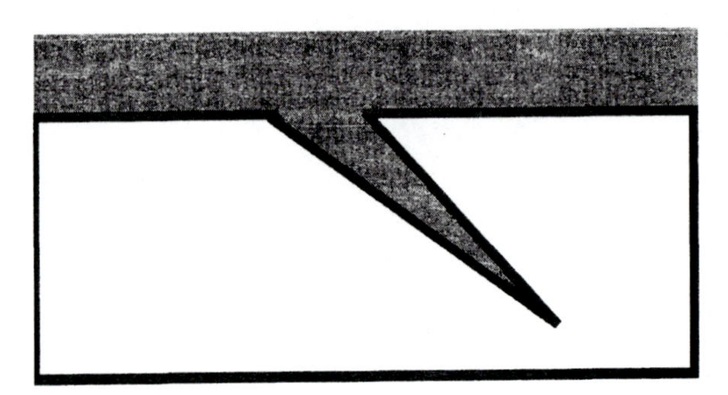

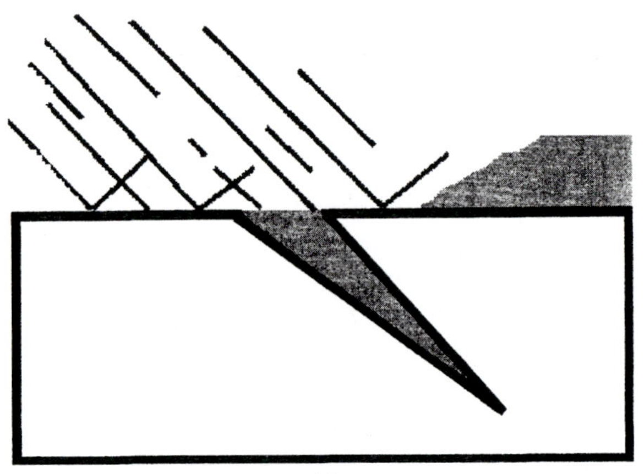

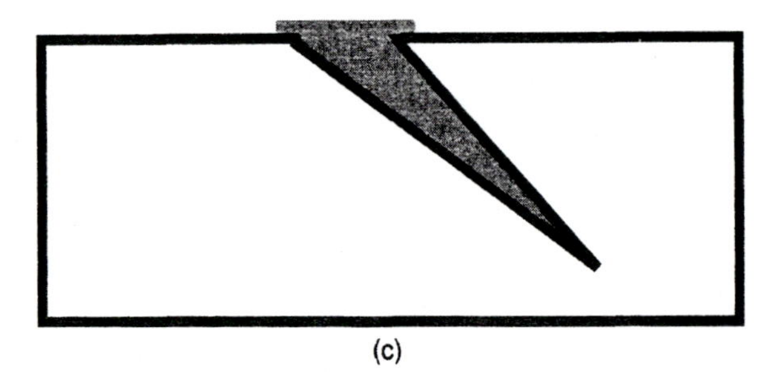

Two basic types of penetrant testing are used today. These are described in the following sections.

Types of Liquid Penetrant Testing

Fluorescent (Type I) Fluorescent penetrants require inspection under a darkened condition using an ultraviolet light called a *black light*. The ultraviolet light enhances or brightens the fluorescent penetrant, allowing better sensitivity and easier detection of flaws. Sensitivity is classified into the following five levels:

Level ½ ultralow
Level 1 low
Level 2 medium
Level 3 high
Level 4 ultrahigh

Visible (Type II) These penetrants are usually bright red or pink dyes that produce good contrast when bleed out occurs against a white developer background. This inspection does not require any power source, black lights, or a darkened area. This method is the most portable and is used for testing small areas. The sensitivity is usual equivalent to a Level 1 sensitivity.

Steps in Liquid Penetrant Inspection

Today, there are six main steps required for a penetrant inspection. They are the following:

1. *Precleaning.* The surface of the part requires some type of cleaning so that the penetrant will soak into the cracks or discontinuities. Cleaning can be as simple as using a solvent cleaner or as detailed as using an etching solution for machined components. After the cleaning stage the part will require some type of drying before applying the penetrant.
2. *Penetrant or dwell time.* After the part has been cleaned, the penetrant is applied to its surface. If only a small area is to be inspected, it may be brushed or sprayed on. If the entire surface requires inspection, dipping is the usual method of application. The penetrant usually is allowed to remain on the part from five to thirty minutes, depending on the procedures used for that particular component.
3. *Removal of excess penetrant.* After the dwell time, the excess penetrant is removed. In the case of a water-washable penetrant, the part is washed under a coarse water spray until the penetrant on its surface is removed. In the case of a postemulsified penetrant, the part is dipped into an emulsifier before the washing stage. Solvent-removable penetrants require some type of solvent to remove the excess penetrant. (These methods will be discussed in more detail later in this chapter.)
4. *Development.* Developers are used to help blot out the penetrant remaining in cracks or discontinuities and to increase the brightness intensity of visible or fluorescent penetrants. The developer can be either wet or dry, depending on the requirements of the procedure. The developer is applied after the part has been cleaned of excess penetrant. A developer is a white powder that leaves a thin film over the part's surface and acts as a blotter. The excess penetrant seeps out of any cracks or crevices and, because of the contrast, makes the defects easier to detect.
5. *Inspection.* After the development time, which ranges from five to fifteen minutes, the part is inspected according to the designated acceptance code. This inspection is performed in good white light conditions when a visible penetrant is used and under darkened conditions with a black light when using a fluorescent penetrant. The black light will enhance the fluorescent penetrant that has seeped back out of any discontinuities, allowing their positions and sizes to be evaluated more easily.

6. *Postcleaning.* Cleaning the part after the inspection process is necessary to remove any remaining penetrant. The chemicals in the penetrant may contaminant the part or react to any chemicals, such as liquid oxygen, that the part contacts.

Penetrant Testing Methods

Four methods of penetrant testing are used:

Method A Water washable
Method B Postemulsifiable lipophilic (oil based)
Method C Solvent removable
Method D Postemulsifiable hydrophilic (water based)

Water-Washable Penetrant *Water-washable penetrant testing (method A)* is probably the easiest and fastest test method. The penetrant is applied to the part by spraying, dipping, or brushing. After a specific dwell time, the excess penetrant is washed off with a coarse water spray. The dwell time will vary, depending on the type of discontinuity that is suspected. The more critical the standard, the longer the dwell time. Longer dwell times allow the penetrant to soak into tight, fine cracks. The disadvantage of water-washable penetrants is that overwashing can wash the penetrant out of any possibly relevant discontinuities.

Solvent-Removable Penetrants *Solvent-removable penetrants (method C)* usually are supplied in a spray can and are used on parts that require a specific area to be inspected (see Figure 11–2). A cleaner is used to clean the testing area; in most cases it is the same cleaner used for removing the excess penetrant and postcleaning. The part's area requiring inspection is sprayed or brushed with a visible bright-red penetrant or a fluorescent one. After the recommended dwell time, the excess penetrant is removed with a lint-free rag. A cleaning solution is applied to the rag for final cleaning, and when residual penetrant is no longer visible on the rag, the part is sprayed with a nonaqueous wet developer. (This developer is wet when applied but quickly dries to a thin, even coat of white powder.) The part is observed for any type of bleed out, which will be evaluated for its size and type. Rejection of the part is based on the accept/reject criteria of the code used. In the case of fluorescent solvent-removable penetrants, the part is observed under a black light. Any bleed out will fluoresce brilliant green.

The main disadvantage of this method is that intricately shaped parts are difficult to clean with a solvent cleaner. Even simple weld reinforcement leaves excessive penetrant remaining. This results in a wide variety of false indications.

Postemulsified Fluorescent Penetrant The *postemulsified fluorescent penetrant (methods B and D)* method is more sensitive in detecting small surface flaws or cracks that may go undetected if the part were overwashed when using a water-washable penetrant. These penetrants are not water washable. After the penetrant has been applied, an intermediate step is needed to make it water washable. This step involves waiting for the appropriate dwell time, after which the parts are dipped into an emulsifier. The emulsifier will mix with the penetrant, making it water soluble

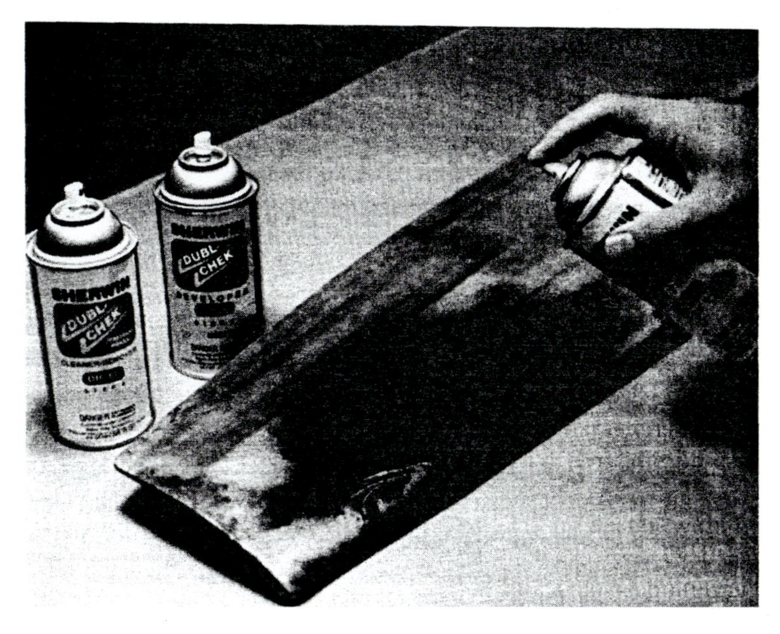

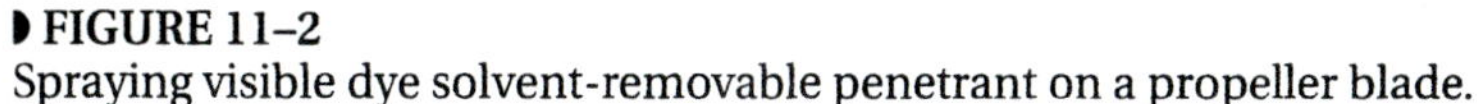

FIGURE 11–2
Spraying visible dye solvent-removable penetrant on a propeller blade.

and allowing the excess penetrant to be washed off. It is critical that the proper emulsification time be used. If the technician allows too little time, the excess penetrant will not be removed, resulting in excessive background fluorescence. If too long an emulsification time is used, the emulsifier will soak into any cracks, resulting in the fluorescent indication being washed away during the rinsing process. The usual emulsification times vary from one to five minutes, depending on the concentration or contamination of the emulsifier. As parts are dipped into the emulsifier, excess penetrant remains in the tank, causing slight contamination. Usually when the time exceeds five minutes, the emulsification is replaced with fresh solution.

Developer

There are four basic types of developers to blot out remaining penetrant:

- Form A—dry developer
- Form B—water soluble
- Form C—water suspendible
- Form D—nonaqueous wet

Dry Developer *Dry developers* are used with a fluorescent penetrant inspection. These are light, fluffy powders that will adhere to the surface of the part, allowing inspection with a black light. They are not used with visible dye penetrants.

Water-Soluble Developer This developer can be used with fluorescent or visible penetrants but is not recommended for use with water-washable penetrants. The developer is a dry white powder mixed with water. It is applied after the rinse stage and before the drying stage. As the part dries, a thin coating of developer remains, allowing for inspection when the part is dry. Because it is water soluble, the developer is removed by a simple water rinse.

Water-Suspendible Developers Water-suspendible developers can be used with fluorescent or visible dye penetrants. They follow the same basic criteria as water-soluble developers.

Nonaqueous Wet Developers Nonaqueous wet developers are commonly used with visible dye penetrants. These developers, supplied in cans, are applied wet. As the developer dries, it leaves a thin coating on the part's surface. After inspection, the developer is removed with a solvent spray.

Advantages and Disadvantages of Liquid Penetrant Testing

The advantages of penetrant testing are that it does not require a highly skilled technician, the inspection is fast, the cost is low, and, in the case of solvent-removable visible dye penetrants, no special electrical power equipment is required.

The disadvantage is that an inspector may have to evaluate a wide variety of false or nonrelevant indications. In the case of water-washable penetrants, overwashing can reduce the ability to interpret small defects. Overdrying the part may dry the penetrant in a small crack, making bleed out difficult or impossible. The parts must be clean to allow the penetrant to soak into any type of surface defect. Contamination may also compromise the sensitivity of the penetrant test.

Ultrasonic Testing (UT)

Ultrasonic testing is a nondestructive testing method that uses high-frequency sound waves to inspect the internal integrity of a wide variety of parts. These may include, but are not limited to, all types of steel, plastics, composites, and forgings. Some types of defects that can be detected are delaminations, disbonds, cracks, shrinkage cavities, bursts, and flakes. Figure 11–3 is an example of a portable ultrasonic machine.

Advantages

The main advantage of UT is its ability to detect flaws deep inside a part. Thicknesses that are routinely tested range from 0.10 inch to 10 feet. Thickness measurements can be accurate to 0.001 inch with specialized equipment. This allows for the detection of extremely small flaws. Ultrasonic testing will determine the flaw's length, width, and height. Another advantage is that UT requires access to only one side of a part. This does limit the accuracy of the inspection, but it is an improvement over radiography, which requires access to two sides of the part.

Most often, ultrasonic testing is used for inspecting metals, especially their welding, brazing, and soldering. Ultrasonic testing also is used extensively in the inspection of electrical components, airframes, composite materials, pressure vessels, bridges, jet engines, pipe wall thicknesses, disbands on satellite components, mining equipment, heavy machinery, and nuclear power plants.

Ultrasonic testing is possible because of the differences in acoustic impedance of various materials. The ultrasonic machine consists of a power supply, transducer, pulser/receiver, and display timer. Other controls that may be included are markers, distance amplitude correction, damping, and gating alarm(s).

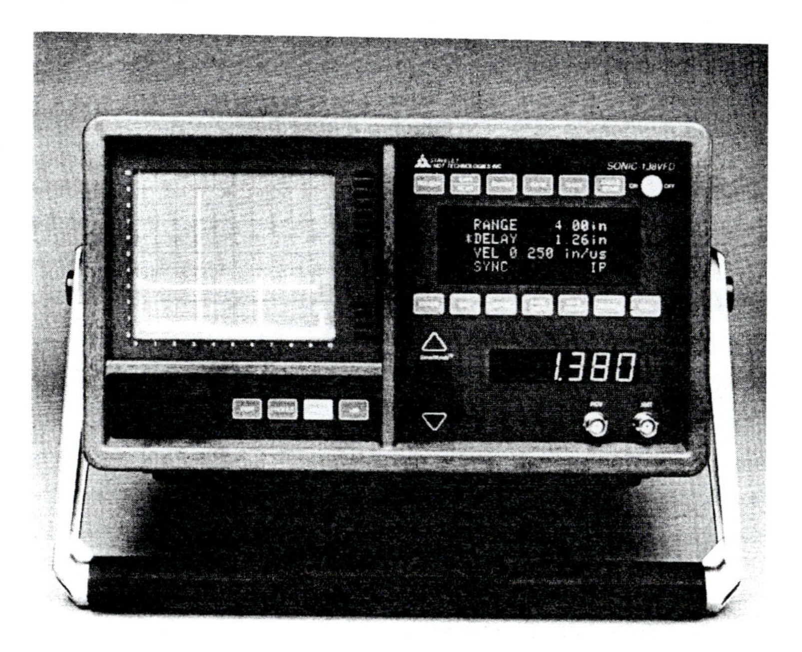

FIGURE 11–3
Sonic 138 VFD ultrasonic machine. *Photo courtesy of Stavely Instruments, Inc.*

The ultrasonic machine creates a burst of electrical energy that is converted to mechanical energy by a transducer. This is called the *piezoelectric effect*. The sound wave produced by the transducer is transmitted through the part until it reflects off the back surface or an area that has a different acoustical impedance, such as air. Air acts as a barrier that reflects the sound energy back to the transducer. This reflection is converted back to electrical energy, and its location is presented on the cathode ray tube (CRT) of the ultrasonic machine. Several hundred readings can be sent and received. The result is a signal (such as the one on the display in Figure 11–4) that represents the depth and, depending on the amplitude of the signal, the approximate size of the defect.

Limitation

One limitation of ultrasonic testing is the requirement of a couplant. A couplant is any material that allows the transmission of sound from the transducer into the part. If the transducer is placed on a dry part, the energy transmitted from the transducer is too small for the ultrasonic machine to amplify to an acceptable signal. This is due to the impedance mismatch between air and the part. A couplant improves the impedance mismatch, allowing sound to be transmitted into the part. It also smooths the surface of the part by filling in any contours that could reflect unwanted sound back to the transducer. A good couplant wets the surface evenly, is easy to apply and remove, is free of air bubbles that would reflect sound energy, is harmless to the part and transducer, and will stay on the test surface for the duration of the

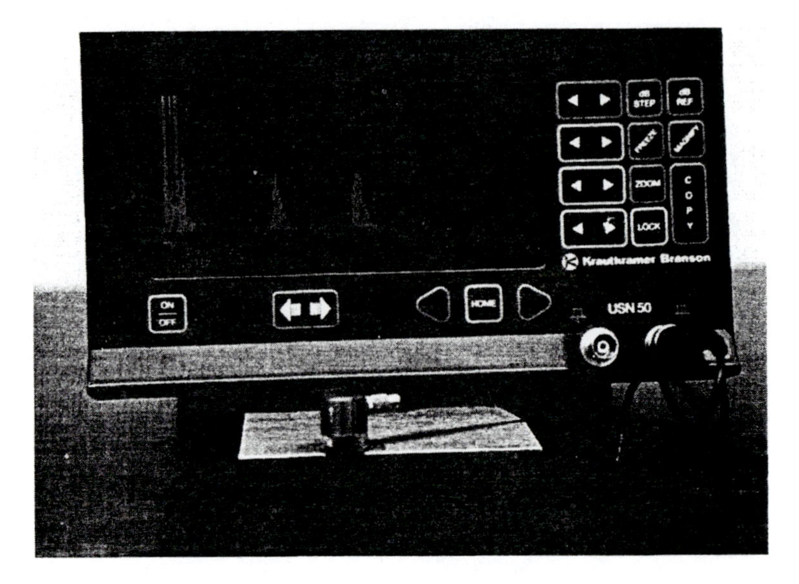

FIGURE 11–4
Ultrasonic machine showing signals from a test block (from left to right, initial pulse, side-drilled hole, and back wall of steel block). *Photo courtesy of Salt Lake Community College.*

testing. When immersion testing is performed, both water with additives and wetting agents are used to prevent air bubbles and possible rusting of the parts.

Transducers

A wide variety of transducers, or search units, is available for ultrasonic testing. They are designed for specific applications, but the basic theory of operation remains the same.

The *piezoelectric transducer* is a plate of polarized ceramic or crystalline material with electrodes attached to its back side that send and receive signals from the ultrasonic machine. Transducers are evaluated based on their sensitivity and resolution.

Sensitivity The sensitivity of an ultrasonic transducer is based on its ability to detect small discontinuities, measured by the amplitude of the signal generated on the CRT. Standard reference blocks are used to determine sensitivity. A known-size flat bottom hole at a specific distance from the transducer is used as a calibration standard. Other standards used are *V* notches and side-drilled holes. It is important to understand that two identical transducers having the same size, frequency, and material will not produce the same level of sensitivity on the ultrasonic screen.

Resolution The resolution, or resolving power, of a transducer refers to its ability to detect two or more flaws located in close proximity to each other, or to detect defects close to the part's surface. When a voltage pulse is supplied to the transducer, the transducer requires time to stop "ringing," or vibrating. Tails, or bursts of sound energy, create a front-surface echo that is wide and high. This signal could mask any small discontinuities located just below the surface of the transducer. Another factor used in determining resolution is the transducer's ability to separate defects in close proximity to each other. If the two defects on the CRT are not distinct, resolution may be unacceptable.

A variety of materials, including quartz, barium titanate, lithium sulfate, lead zirconate, and tourmaline exhibits piezoelectric properties. The three most common piezoelectric materials that are used in transducers are quartz, lithium sulfate, and polarized ceramics.

Quartz is best known for its thermal, chemical, and electrical stability. It is very stable, does not change over time, and has excellent strength. Its major limitation is its inability to convert mechanical energy efficiently, which makes it the least efficient generator of acoustic energy.

Lithium sulfate transducers are the most efficient receivers of ultrasonic energy and are used as the receiver transducer in through-wave transmission applications. They are average transmitters, do not age well, are soluble in water, and are limited to temperatures below 165°F.

Ceramic or polarized ceramic transducers are the most efficient transmitters of sound and are used in through-transmission testing. They are unaffected by water and can be used at high temperatures. Limitations include their low mechanical strength and a tendency to age.

Transducer Types

A transducer's size and shape are limited only by the imagination of the engineers. Their sizes range from ¼ inch to as wide as 6-inch paintbrush transducers. The application and the need to detect a specific size defect determine which transducer is used. Larger transducers are used when less beam spread is required; lower frequencies usually are used with large transducers, and more penetration can be obtained with larger transducer of the same frequency. This reduces scanning time; however, sensitivity is compromised.

A smaller transducer will increase the sensitivity of the test, but less penetration and more beam spread are disadvantages. Also, more time is required to inspect the part because of the small scanning area of the transducer.

Contact Transducers *Contact transducers* are used in both straight and angle beam testing. These transducers are used to inspect parts where the transducer is slid across the surface of the part by the technician. A liquid medium or couplant such as oil or glycerin is used as a medium between the transducer and the part. A wear ring with an aluminum oxide coating is incorporated to reduce transducer wear. In some instances a plastic wedge is attached, which improves near-surface resolution when testing thin materials.

Angle Beam Transducers *Angle beam transducers* are straight beam transducers mounted on a plastic or Lucite wedge (an example is shown in Figure 11–5). The wedge directs the sound energy into the part at an angle; the sound waves created can be shear, longitudinal, surface, and plate waves. When shear waves are required, the wedges used most commonly induce a refracted angle of 45, 60, or 70°. Angle beam contact testing is used to locate discontinuities orientated between 90 and 180° to the surface and is often used for the inspection of welds. The weld reinforcement prevents

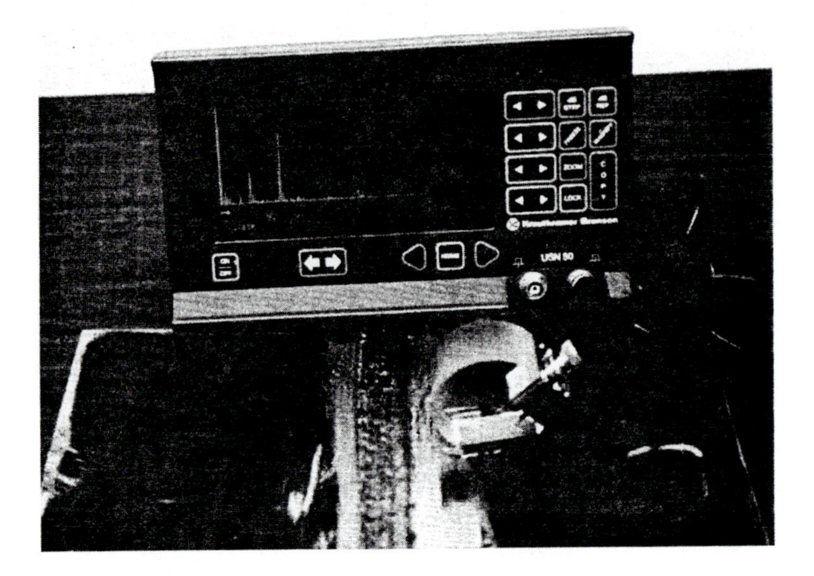

FIGURE 11–5
Ultrasonic machine showing angle beam inspection of a weld plate. *Photo courtesy of Salt Lake Community College.*

ultrasonic inspection over the surface of the weld; therefore, angle beam testing is the most acceptable procedure.

Focused Transducers Focused transducers are designed to concentrate sound energy. Focusing the sound beam has advantages and disadvantages. These transducers are used to obtain maximum sensitivity in a small area. The maximum sound energy is moved toward the transducer, improving its near-surface resolution but limiting the usable range of the transducer. Focusing also can reduce the effects of a rough surface and metal noise.

Paintbrush Transducers *Paintbrush transducers* are approximately 6-inch transducers made up of a mosaic pattern of smaller crystals. These crystals are matched so that the intensity of the sound energy is consistent over the length of the transducer. They are used to scan large areas of a part and can locate discontinuities quickly. When a discontinuity is found, smaller, more sensitive transducers are used to size and evaluate it.

Frequency

In ultrasonic testing the proper frequency of the transducer is always a compromise. Transducers vary in frequency from 1.0Mhz to 50Mhz. Sensitivity, penetration, beam spread, and resolution are all factors when determining the proper frequency of a particular test.

The higher the transducer's frequency, the smaller the defect that can be located. The smallest defect that can be detected is one-half the wavelength of the sound energy. The wavelength is determined by this simple formula:

Wavelength = Velocity/Frequency.

The *velocity* is the speed at which sound will travel through a particular material: Each particular material allows sound to travel through it at a specific speed. Higher frequencies create shorter wavelengths, allowing smaller defects to be located.

The disadvantage of higher frequencies is that the sound will not travel as far into the part when compared to lower frequencies. If a technician were inspecting thick parts, he or she would need to use a lower frequency, therefore compromising sensitivity.

Penetration, the maximum depth at which a discontinuity can be detected, can be increased by using a lower frequency. Another variable that affects the depth of ultrasonic waves is the grain structure of the material. Coarse material, such as castings or composite material, reduces the sensitivity of high-frequency transducers. Lower frequencies allow more penetration.

Beam spread is the ability of sound to spread or diverge from the face or central axis of the transducer. As the frequency increases, the beam spread decreases, allowing for more accurate testing results. The size of the transducer can also affect beam spread. Larger transducers have lower beam spreads than smaller transducers of the same frequency.

Disadvantages

A disadvantage of ultrasonic testing is that the equipment can be expensive, running from approximately $5,000 for a portable ultrasonic machine to over $100,000 for an automated system.

There are other drawbacks. Technicians need extensive training and technical knowledge to obtain accurate results. When contact testing is used, the parts need to have a fairly smooth surface. (A rough surface would interfere with the sound energy leaving and entering the transducer.) In some situations, the flaw may be near the surface of the part and might not be detected due to an initial pulse or dead zone associated with ultrasonic transducers. There is also a need for some type of couplant.

Eddy Current Testing

Eddy current testing is based on electromagnetic induction. In 1831, Michael Faraday discovered that a current-carrying wire would induce a small amount of current in an adjacent coil of wire only when the current in the primary coil was turned on or off. From this he determined that the changing magnetic field induced the small amount of current. Later experiments confirmed Faraday's conclusion that by placing a current-carrying coil near a second coil, a current was induced into the second coil. This would occur only when the current in the primary coil was turned on or off. It was concluded that the magnetic field created this current. Eddy currents simply use one coil of wire to induce a small current into the part under the coil. Using AC power, the current follows a sine wave traveling from positive to negative, the equivalent of the current being turned on and off in Faraday's original discovery. Eddy currents can therefore be induced into any conductive part.

Electromagnetic induction allows a wide variety of inspections to be performed using an eddy current machine. The following are some types of inspection:

- Measuring the thickness of nonconductive coatings. This could be a measurement of paint, adhesives, or plastics. Cladding and plating thicknesses also can be inspected.
- Examining and identifying the electrical conductivity of parts. Conductivity can change when the metal is heated, placed under loading, chemically treated, and shaped.
- Sorting dissimilar metals based on heat treatment, composition, or microstructure changes during manufacturing.
- Performing high-speed testing in automated systems, resulting in a fast and reliable inspection. Because of electromagnetic induction, the part does not need to be in direct contact with the probe or coil.
- Using an internal or external coil to test the entire circumference of the part or a small probe to inspect critical areas that have irregular dimensions.
- Determining crack size and locations (see Figure 11–6).

Basic Operation

A coil that is classified as a surface, encircling, or internal bobbin coil is placed next to the part. An alternating current runs through the coil and is designated the *exciting current.* This current causes eddy currents to flow in the part, allowing inspection. Eddy currents have a range of penetration based on the frequency of the

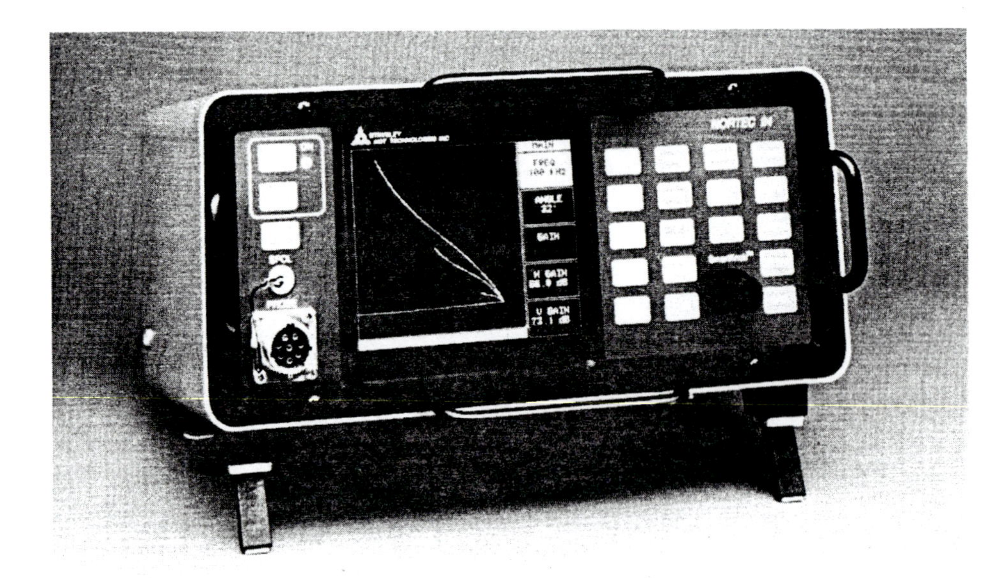

FIGURE 11–6
Nortec 24 impedance plane machine showing signals obtained from three different crack sizes. *Photo courtesy of Stavely Instruments, Inc.*

coil and the type of material under inspection. Depth of penetration can reach 5 inches using a 100kHz probe on stainless steel, or 0.001 inches using a 100kHz probe on ingot iron. In most cases, testing is limited to defects on or close to the part's surface.

Eddy currents flow in closed loops in the part. There is a correlation between magnitude and timing that can be changed by changing the original or primary field generated by the coil. The conductivity of the part can also affect eddy currents since a current is a variable.

If the conductivity of the part changes due to improper heat treatment, this will affect the eddy currents flowing into it. A crack will alter the shape of the eddy currents and they will no longer be able to travel in a circular path. Instead, they will try to continue their pattern around the crack. These electromagnetic changes can be interpreted and evaluated by an experienced technician, and the part can be accepted or rejected based on the codes used.

The technician is always making a compromise among penetration, sensitivity, and speed of the inspection. These variables, and those associated with the type of part under inspection, can involve the following:

- Coil impedance
- Electrical conductivity
- Magnetic permeability
- Liftoff
- Edge effect
- Skin effect
- Fill factor

Coil Impedance When alternating current flows through a coil, two limitations are present: the AC resistance of the wire, designated as R, and the quantity of *inductive reactance* in ohms, which is designated as X_L. (Inductive reactance is the combined effect of test frequency and coil inductance.) There is also the effect of impedance, Z. This is the total resistance to the flow of alternating current and can be expressed as

$$Z = \sqrt{R^2 + {X_L}^2}$$

Figure 11–7 shows the signals generated by different materials when using an impedance plane machine.

Electrical Conductivity Every type of material allows current to pass through it at different levels. A material's ability to conduct current can be measured in either *conductivity* or *resistivity*, based on the number of amperes of current that will flow through a given size of the material when a given voltage is applied to the material. Pure annealed copper is used as a standard of 100 percent. All other materials are expressed as a percentage of this standard. This is called the *International Annealed Copper Standard (IACS)*. For example, magnesium can carry only 37 percent of the current that can be carried by the same size copper wire at a given voltage.

Copper, which has a low resistance to the flow of electricity, is classified as a *conductor*. Materials that have an intermediate resistance to current flow are called

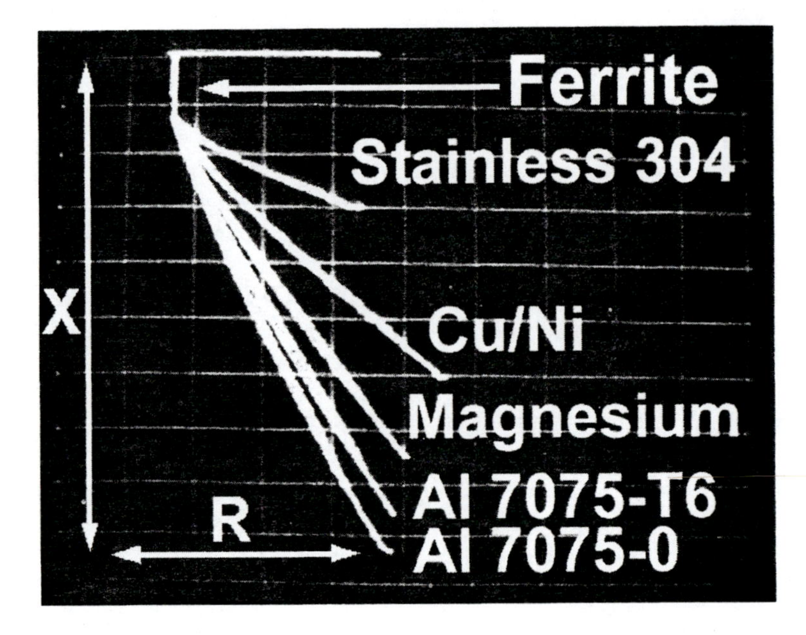

FIGURE 11–7
Eddy current impedance plane machine showing signals generated from different materials.

semiconductors. Materials that have a high resistance to the flow of electricity are called *insulators*.

Any change in conductivity can be detected and evaluated with eddy current testing. A variety of factors can change a part's conductivity. *Alloys* are a combination of two or more elements, or of a metal and some other type of base material. Because each metal has its own chemical characteristics, the combination of elements will affect the conductivity of the final product. These material changes can be detected using eddy current testing.

The hardness of a part can change when the metal or alloy is subjected to heat treatment. Depending on the material and the heat-treating process, the metal may become harder or softer.

Temperature can affect the conductivity of a part as well. An increase in temperature normally results in a decrease in the part's conductivity. Residual stresses can also affect the eddy current results; they are more unpredictable.

Magnetic Permeability Ferromagnetic materials such as iron and nickel have the ability to retain a magnetic field. This magnetic field can greatly affect the results of an eddy current inspection. When the eddy current probe is placed on a nonmagnetized ferromagnetic material, the magnetic field generated by the probe is intensified by the magnetic properties of the part. This variable can result in inconsistencies when conducting eddy current testing on ferromagnetic materials. The same factors that change the electrical conductivity of a part may also influence its magnetic permeability. To reduce this inconsistency, the part needs to be magnetically saturated so that the magnetic variable is eliminated. After testing, the part requires demagnetization.

Liftoff Eddy current probes are very sensitive when there is a liftoff, that is, a separation between the probe and the part. This can be an advantage or a disadvantage, depending on the type of inspection the technician is trying to perform.

Small liftoff changes can be detected, so nonconductive coating thicknesses such as paint can be evaluated with excellent accuracy. In the case of paint thickness inspection, the technician is not concerned with any type of flaw in the conducting material, but is looking at the signal's position in relation to the paint thickness.

Figure 11–8 shows a variety of probes used in eddy current testing.

Edge Effect The eddy currents generated by the coil or probe will always travel in circular paths. As discussed earlier, if these paths are changed (for example, by a crack), the signal on the eddy current machine will change. This allows the technician to identify any discontinuities in the part. The major limitation of eddy current testing occurs at the edges of parts. The eddy currents cannot remain in a consistent circular pattern and are distorted by the edge, creating a condition known as *edge effect.* This can mask any relevant flaw or show a false indication. Some probes are designed to help eliminate edge effect by shaping the eddy current pattern.

Skin Effect Both the thickness and shape of a part will affect eddy current responses. Eddy currents are distributed evenly through the thickness of the part under inspection. They are strongest on the surface of the part next to the probe or coil and become weaker with increasing depth. This is known as *skin effect.* Because of this decreasing effect, there will always be some point where no eddy currents are generated. To determine the depth at which flaws can be detected, the standard depth of penetration is calculated at 37 percent of the density of the eddy currents on the surface. This depth will change, depending on the conductivity of the part and its magnetic permeability. If the conductivity, permeability, or probe frequency increases, the depth of the eddy currents will decrease. Defects lower than 37 percent are difficult to identify, resulting in false or inaccurate information.

FIGURE 11–8
Various eddy current surface probes.

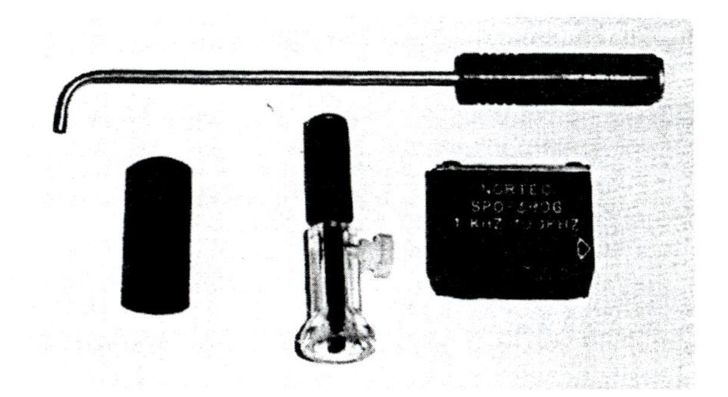

If thin material is inspected, the eddy currents may penetrate deeper than the thickness of the part. As long as the part thickness does not change, consistent testing can be performed. If the thickness begins to vary, a false reading may occur. The thickness of the part should be two to three times the standard depth of penetration. If the technician is looking for thickness changes in thin material, eddy current testing allows these changes to be identified easily.

Fill Factor The *fill factor* of an encircling coil is similar to liftoff of a surface probe. When using an encircling coil, the part's outside diameter is compared to the coil's inside diameter. The shorter the distance between the two, the better the fill factor. A good fill factor is needed to induce eddy currents into the part. The part is positioned in the center of the coil to obtain eddy currents evenly throughout the part's surface. If the part did not remain centered, a false reading would be detected. The main advantage to using an encircling coil is that the entire part can be inspected at a high rate of speed. Inspecting rod and bar stock is commonly associated with encircling coils. Because eddy currents are sensitive to positioning, they can detect changes in the diameter of parts.

Disadvantages

As stated earlier, eddy current penetration is limited. Parts under inspection are required to be conductive, and ferromagnetic parts may be difficult to inspect. Equipment costs can be high and initial calibrations can be complex.

Radiographic Testing

Radiography is the process of using short-wavelength electromagnetic radiation to detect internal flaws in a part using X rays or radioactive isotopes. A two-dimensional latent image from a source of radiation is produced on a sheet of film, which can be developed manually or automatically.

X-ray tubes are electronic devices that convert electrical current into X rays. The basic X-ray tube is an enclosed high-vacuum envelope of Pyrex®-type glass that has a high melting point. All electrical connections are made to withstand the high temperature generated by the tube. These tubes come in many shapes and sizes, depending on the type of X-ray machine. Figure 11–9 shows a portable X-ray machine.

Inside are two electrodes, called the *cathode* and *anode*. The cathode is made up of a focusing cup and filament that, when heated, is a source of free electrons. The focusing cup is constructed of pure iron and nickel and helps direct the electrons toward the anode. The filament is constructed of tungsten and emits free electrons when the electrical current flows through it. The milliamperage control on the X-ray machine adjusts the flow of current, thus regulating the amount of electrons emitted. These electrons have a negative charge.

The target, or anode, is made from tungsten, gold, or platinum. A backing material of copper is added to improve cooling. The anode is made up of a dense material to obtain the maximum number of collisions between the electrons generated by the cathode. Also, the material is required to have a high melting point because of the high temperatures generated. The main disadvantage of X-ray machines is the generation of heat. The process of generating X rays is very inefficient. Approximately

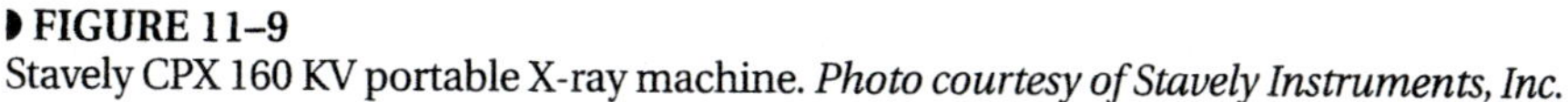

FIGURE 11–9
Stavely CPX 160 KV portable X-ray machine. *Photo courtesy of Stavely Instruments, Inc.*

98 percent of the energy is expended as heat. This heat is removed from the anode by a variety of methods, such as copper cooling fins, oil, water, gas, and injection cooling.

X rays are produced from the collision of fast-moving electrons striking a target made of dense material such as tungsten. Two types of X rays are produced. When the electrons strike the target material, they lose their speed as they interact with the target. These X rays have a wide variety of wavelengths and are referred to as *continuous X rays*. The second type of X ray, *characteristic X rays*, is created when an electron collides with an atom of the target. This causes a transition of the electron orbiting the atom. As the atoms try to rearrange themselves, X rays are emitted with a more concentrated wavelength. This type of X ray has higher intensity.

Particle Radiation

Ionizing radiation exists as particles and waves in the electromagnetic spectrum. Moving at nearly the speed of light, negative-charged *beta particles* can be stopped by a thin sheet of paper or several millimeters of skin. Positive-charged *alpha particles* are unable to penetrate a thin sheet of paper and can be stopped by the surface of the skin. Both alpha and beta particles are very dangerous to the human body if they are ingested or inhaled.

Gamma Rays

Gamma rays travel at the speed of light and have no electrical charge. They have great penetrating power, allowing a wide variety of parts to be examined, and can pass easily through the human body, damaging living tissue. If excessive damage takes place, the body cannot recover, resulting in permanent injury or death. Three radioactive isotopes that produce gamma rays are radium, cobalt 60, and iridium 192.

Radioisotope Testing

A variety of leak test methods, called *radioisotope testing*, involves the use of radioactive isotopes. The isotopes are tracers added to a base solution, such as water or hydrocarbons. The most common isotope added to water is the sodium isotope 24 Na. (This isotope has a half-life of only 15 hours and can be stored in the vessel until its activity or radiation level is below the acceptable disposal standard; this standard is usually below the drinking tolerance.) A vessel is filled with sodium bicarbonate solution and pressurized to the required limit. After the vessel has held pressure, it is drained and cleaned with fresh water. A special type of Geiger counter is then used to detect the presence of any leaks in the vessel.

These isotopes emit different energy levels of radiation in inspecting components or calibration. They are separate in radiography only because they are not generated by a source of power such as an X-ray machine.

Radium A natural source of radiation is uranium. After the discovery of X rays, the scientist Henri Becquerel discovered radiation emitting from uranium-bearing material. This material was later refined into a more concentrated substance by Marie and Pierre Curie. They named the element *radium*. Radium has a half-life of approximately 1,600 years. Through decomposition, radium produces radon and other radioactive daughter products. The disintegration of these daughter products emits gamma rays. Pure radium consists of radium sulfate, and it is now packaged in capsules. With its long half-life, it is sometimes used as a calibration standard.

Cobalt 60 *Cobalt 60 (Co60)* is a man-made radioactive isotope created when cobalt is bombarded with neutrons. It has a half-life of 5.27 years. The primary gamma ray emission consists of 1.17 to 1.33 million electron volts (mev). This number represents the energy equivalent in comparison to an X-ray machine. The radioactive isotope is supplied as a capsule, and is transported in a lead and depleted-uranium container (shown in Figure 11–10) weighing approximately 500 pounds. Cobalt 60 is used primarily for the radiography of large castings and weldments ranging in thickness from 1 to 8 inches. It also is used extensively in industrial radiography. Its main disadvantage is the result of its penetration capabilities: Extensive shielding of the technician is required.

Iridium 192 *Iridium 192 (Ir192)* is another man-made radioactive isotope created by neutron bombardment. It has a half-life of 74.3 days. Its primary gamma-ray emissions are 0.31, 0.47, and 0.60 mev. It does not have the penetrating ability of Co 60; therefore, it is more portable. Figure 11–11 shows the container in which iridium

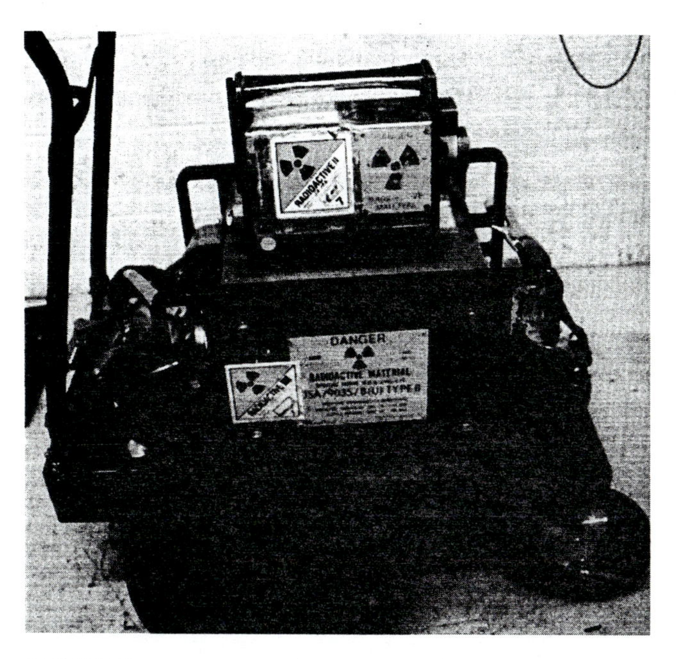

FIGURE 11–10
IR192 camera atop a Co60 camera. *Photo courtesy of Met-Chem Testing Laboratory.*

FIGURE 11–11
IR192 camera. *Photo courtesy of Met-Chem Testing Laboratory.*

192 is stored (often called a *camera*) is approximately 40 pounds and allows for much easier transport. Ir192 is used for testing steel components up to 2.5 inches.

Advantages of Radiography

Radiography is most often used to detect internal defects, although it can also identify surface defects. The image of the part is produced on film, which is an excellent permanent record of both acceptable and rejectable parts.

Radiography can detect defects that are parallel to the source of radiation, as well as large defects. The ability to detect planar defects such as cracks or laminations is dependent on its orientation. Defects that have a thickness of approximately 1 percent of the total part thickness can be detected. Identifying and interpreting smaller defects, is difficult, if not impossible.

A wide variety of materials can be inspected using radiography. These include, but are not limited to, steel, aluminum, circuit boards, valve assemblies, plastics, explosives, concrete, and electronic components.

Radiography is often used for inspecting weldments and castings. This ensures that the products are free of internal defects that could cause a premature failure. Equipment and parts that are extensively tested include assemblies for steam-power equipment, pipelines, and related monitoring equipment in the petrochemical industry, high-pressure systems, turbine generators, and pressure vessels.

Radiography is also well suited for inspecting electrical and electronic components. It can easily detect voids in the mount and seal areas of semiconductors. Any type of cracks, broken or damaged wires, foreign material, or voids can be detected.

Limitations of Radiography

The equipment required to handle and store radioactive isotopes will range from \$6,000 to \$15,000. The cost of the radioactive isotope is approximately \$2,000 for IR192 to \$10,000 for Co60. X-ray machines range in price from \$8,000 to \$80,000, with fully automate systems exceeding \$500,000.

Radiography cannot detect all defects. Tight cracks less than 1 percent of the total part thickness often can go undetected. The orientation of the source of radiation and the crack should be as close to parallel as possible. If not, there may not be enough thickness variation for the crack to show up as a dark indication on the film. Minute flaws such as flakes, micro-porosity, laminations, and inclusions also can be difficult to detect.

It is very important to adhere to proper safety standards. Radiation can be very dangerous, and the proper safety training and monitoring equipment are necessary. In most accidents involving industrial radiation, the operator did not monitor the radiation area. This is the leading cause of unnecessary overexposure.

Radiation Damage to Humans Radiation is measured by its intensity and the amount of it that is absorbed as it passes through 1 cubic centimeter of dry air. The unit of measurement designated as *roentgen per hour* is the product of the intensity and the absorption coefficient for air.

An individual's exposure to gamma radiation is measured in either roentgens (R) or sieverts (Sv). The amount of radiation an individual is exposed to is called *roentgen equivalent in man (rem)*. Because exposures are usually very small, the term *millirem* is used; it is 1/1000th of a rem. The Nuclear Regulatory Commission (NRC) has set acceptable limits for individuals who work in radiography. A person can receive a maximum of 1.25 rem (1,250 millirem) every three months, or a maximum of 5 rem (5,000 millirem) per year.

To put this in better perspective, the following table lists the effects of radiation based on whole-body exposure within a 24-hour period.

Amount	Result
0–25 rem	No injury; first detectable blood change at 5 rem
25–50 rem	Blood changes, no serious injury
100–200 rem	Injury with possible disability
200–400 rem	Injury and disability certain, death possible
400–500 rem	Median lethal dose (MLD); 50% fatalities
500–1,000 rem	Up to 100% fatalities
Over 1,000 rem	100% fatalities

Radiation Safety and Detection Devices

All individuals performing radiography and the radiography facilities are monitored for radiation exposure. The facilities are equipped with calibrated instruments that can detect excessive radiation levels, alerting personnel of a potential radiation hazard. Periodic monitoring of the perimeter with portable monitoring equipment is also performed. When X-ray machines are used, an interlock and in some cases cameras are used to secure the radiation area. The interlock will disengage all power to the X-ray machine if all shielding doors are not closed. A ten- to thirty-second audible and visual alarm is activated before the X-ray machine engages. This allows any technician an opportunity to disengage the power before he or she is exposed to radiation.

All individuals are also monitored with dosimeters, film badges, Geiger counters, or survey meters, which are shown in Figure 11–12.

Dosimeter A *pocket dosimeter* is an air-filled ion chamber that operates under the principle that like or similar charges repel both each other and radiation-caused ionization in a gas. They are approximately 4 inches long and ½ inch in width, and provide an immediate measurement of the radiation dose. The essential parts are the quartz-fiber electrode consisting of fixed and movable sections, a transparent scale, and a lens. An external source is used to charge the dosimeter, moving the quartz fiber to the zero position on the scale. When the dosimeter is exposed to ionizing radiation, the ions created will neutralize the charge on the wire. The fiber moves toward the fixed position as the dosimeter discharges. The position is a direct measure of radiation and allows the technician to verify his or her exposure.

Film Badge A *film badge* consists of a small piece of film that's placed in a film holder. The film holder records the types of radiation that the individual is exposed to. Film badges are replaced every month and the film is developed using standard techniques. The density of the film is measured and can determine the amount of radiation that the technician received. A monthly report on each individual is sent to the company. This allows the radiation safety officer to keep track of yearly and quarterly totals. A technician receiving close to his or her maximum limits can be reassigned.

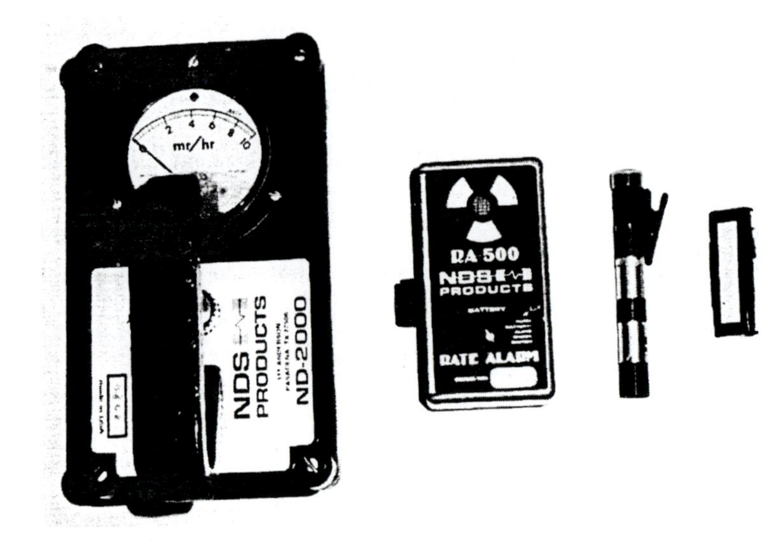

FIGURE 11–12
From left to right, a Geiger counter, alarm ratemeter, dosimeter, and film badge.

Geiger Counter *Geiger counters* (survey meters) utilize a Geiger-Muller tube as an ionization chamber. The voltage differences between the anode and the cathode and the gas inside the tube create a reading on the survey meter. This reading can be audible, visual light, digital, or meter deflection. Survey meters require calibration every three months. The disadvantage of Geiger counters is their tendency to block out excessively high radiation levels, showing a false reading of zero. If high radiation levels are expected, another type of survey meter should be used.

Area Alarm System Area alarm systems are usually installed in permanent facilities. They are calibrated to emit a visual and audible alarm when a specific level of radiation is exceeded. When Co60 or Ir192 is used, these alarms will operate at 2 milliroentgens per hour.

Magnetic Particle Testing

Magnetic particle testing, the last nondestructive testing method covered in this chapter, has the ability to reveal surface and slightly subsurface defects on components and materials that can be magnetized. These are called *ferromagnetic materials*. The testing method is based on the principle that any defect in the part will distort the magnetic field, creating flux leakage.

Magnetic particle testing is used most often as a final inspection and for in-service quality control. Final inspection is needed to ensure that all defects resulting from the manufacturing process are located. In-service inspection can be performed during routine maintenance. Because of cyclic or static loading, critical parts are tested to prevent catastrophic failures. Routine magnetic particle testing is performed on flywheels, crankshafts, aircraft components, railroad parts, crane hooks, turbines, blades, weldments, and pressure vessels.

Magnetic particle testing is used mainly to locate surface discontinuities, including fatigue cracks, seams, laps, quenching and grinding cracks, and surface ruptures in castings, forgings, and weldments. It is an inexpensive testing method that requires minimal training. Identified discontinuities can be interpreted directly. No electronic evaluation or interpretation is required, as with ultrasonic or eddy current testing. When used properly, magnetic particle testing has excellent sensitivity in locating small cracks and surface defects. Virtually any size part can be tested. It has limited sensitivity to large cracks and subsurface discontinuities, due to the type of flux leakage created by a wide opening and the magnetic field created by the equipment. In some cases, subsurface defects can be located. If the discontinuity is "fine sharp" and close to the surface, and the magnetic field is transverse to the indication, it most likely will be located.

To detect these types of indications, there must be a field of sufficient strength in the proper orientation that will create enough flux leakage to be seen. The discontinuity will produce the strongest possible flux leakage when it is oriented at right angles to the magnetic field (see Figure 11–13).

The induced magnetic field must then interact with a discontinuity to form an indication. The magnetic lines of force must create a bridge across the discontinuity, creating flux leakage. For the best results, these lines of force should be parallel to the discontinuity. Since the location of any flaw is unknown, parts are inspected in two directions 90° to each other.

Magnetizing Current

Alternating current (AC) and direct current (DC) are both used when testing with magnetic particles. Each has its own advantages and disadvantages.

Alternating Current Alternating current is used when the surface of the part requires maximum sensitivity. The rapid reversal of the magnetic fields induced by alternating current agitates the particles. As the particles become attracted to

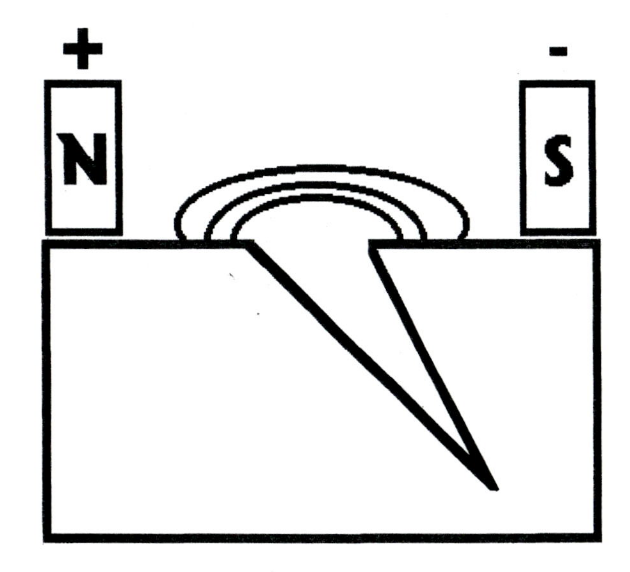

FIGURE 11–13
Flux leakage created by a crack when a magnetic field is applied; *N* denotes north, and *S* denotes south.

the flux leakage of a discontinuity, they accumulate at the leakage field, resulting in a stronger visual indication. This helps the technician locate the indication more easily. Alternating currents of 110, 220, or 440 volts can be used. The current usually has a frequency of 50 or 60Hz and is stepped down to lower voltages with transformers. (Lower voltages allow higher amperages; these can be up to several thousand amps.)

One disadvantage of using alternation current is that demagnetization of the part may be a potential problem. If the magnetic field is stopped at the peak of the alternating current cycle, a large residual magnetic field may be induced into the part. This may be difficult to remove.

Direct Current Direct current is used when subsurface defects are suspected because it has the ability to penetrate the surface. Direct current is obtained by rectifying alternating current. Rectifiers are used to produce a straight direct current or, in the case of single-phase alternating current, the current is permitted to flow in only one direction. Direct current's disadvantage is that it reduces the sensitivity of surface discontinuities. Still, this is a good testing method when inspecting castings or weldments.

To perform an accurate magnetic particle inspection, you need to properly magnetize the part. Any type of indication creates flux leakage, so it is important that the magnetic fields are perpendicular to the field.

Equipment

Electromagnetic Yokes An electromagnetic yoke is a small, portable machine that can be used in either the AC or DC mode. Essentially, it is a small coil wound around an iron core. The magnetic field produced by the yoke induces a longitudinal field in the part being inspected. It is important to place the area of interest directly between the legs of the yoke to obtain maximum sensitivity. Figure 11–14 shows an electromagnetic yoke.

Coils Most coils used in magnetic particle testing are either in a fixed frame or are flexible cables that can be wrapped around the part under inspection. In most cases, the effective magnetic field is only 6 to 9 inches on either side of the coil. When longer parts are inspected, either the part must be moved through the coil or the coil must be moved along the length of the part. The proper current is required when using coils. This is based on the length-to-diameter (L/D) ratio. The number of ampere turns required for a proper magnetizing field is given by the following equation:

$$NI = 45{,}000\ (L/D)$$

where N represents the number of turns in the coil, I represents amps, and 45,000 is a constant.

Central Conductor When inspecting tubing and ring-shaped parts, a *central conductor* is used to induce a magnetic field into the part. This conductor is made of solid or hollow tubular material with good conducting capabilities. Copper rod of various sizes is typically used. The rod is inserted through the inside diameter of the part and the current is generated through the central conductor, inducing a magnetizing field into the part. Figure 11–15 shows a magnetic particle inspection using a

FIGURE 11–14
An electromagnetic yoke used to inspect a weld part.

central conductor. One of the disadvantages of using a central conductor is that the magnetic field it generates is limited to four times the diameter of the conductor. As a result, large circular parts may require multiple magnetizations.

Direct Contact If the part under inspection does not have any holes or openings, *direct contact* may be used. The part is clamped between two contact points and a current is generated through it. This is referred to as a *head shot*. A major disadvantage of using a head shot is the use of the contact points. Copper pads are used to improve both the use of contact area and the current flow. The part is locked in place by pressure to prevent arcing. Any type of arcing or excessive pressure could damage the part being inspected.

Prods *Prods* are used to inspect large parts. Two copper rods are spaced 4 to 8 inches apart and connected to a power source. When placed on the surface of the part, they generate a circular magnetic field. Prod inspection often is used for inspecting the root and hot pass of welds and castings. Its portability allows for on-site inspection. Major limitations include excess arcing and heating.

Magnetic Particles There is a variety of magnetic particles available, classified according to the medium that is used as a carrier, such as air for dry magnetic particles, and liquid when wet particles are used. The particles have to meet specific characteristics: size, shape, mobility, visibility, density, and contrast.

FIGURE 11–15
A copper rod used as a central conductor for magnetic particle inspection. *Photo courtesy of Salt Lake Community College.*

If the particles used are too heavy, the magnetic field may not be strong enough to retain them where the flux leakage was created. On the other hand, if the particles are too fine, they may adhere to the surface in areas where there are no discontinuities.

The properties of the magnetic particles should include high magnetic permeability, low coercive force, and low retentivity. The particles should have a long and slender shape. These types of particles will develop significant north and south poles that help the particles to arrange themselves in strings when flux leakage occurs.

Dry particles come in a variety of colors to obtain maximum contrast between the particles and the part. These include black, yellow, silver, red, and blue. They should have the ability to drift freely to the surface of the part, creating a uniform coverage. This allows the particles to be attracted where the flux leakage has been created from any type of discontinuity. Dry particles are most sensitive when inspections are performed on rough surfaces. When used with direct current, they are more effective in detecting subsurface defects than wet particles.

Wet particles are suspended in either an oil- or water-based carrier. Red and black are available, but fluorescent particles are used most often. They move more slowly toward any type of flux leakage because of the carrier. The particles are very fine and tend to clump together. This clumping creates disproportional shapes that tend to line up with any leakage fields, an effect that helps increase the sensitivity of wet fluorescent particles. Wet particles are better than dry particles in detecting fine surface discontinuities. This sensitivity can be increased by using alternating current. Wet particles also are used with direct current to detect slight subsurface discontinuities.

Limitations

Any part tested using magnetic particles must be ferromagnetic. Testing is used mainly to detect surface discontinuities, with limited use in detecting subsurface flaws. Inspection of parts is required in two directions. After inspection, cleaning and removing magnetic particles may be difficult. Large currents may be required to induce the proper magnetic field, and this may produce arcing or burning.

Another potential problem is magnetic writing. Magnetic writing is usually created on parts with high residual magnetism characteristics. If a part is contacted with the sharp edge of another part, a residual magnetic field is created. This results in flux leakage that requires interpretation.

QUESTIONS FOR REVIEW

1. How does contamination affect a leak test?
2. What is the disadvantage of using a leak test?
3. What affects the wetting ability of penetrants?
4. When performing a penetrant examination, what is the purpose of the developer?
5. What is the purpose of a couplant?
6. How does resolution affect the interpretation of ultrasonic signals?
7. What is the advantage of using a higher frequency transducer?
8. What type of material is required of the test part to perform an eddy current examination?
9. When can ferromagnetic parts undergo inspection using eddy current testing?
10. What two types of X rays are produced from an X-ray machine?
11. What type of materials can be inspected when performing a magnetic particle examination?
12. What are three disadvantages of magnetic particle testing?
13. What magnetic properties are important when using dry magnetic particles?

ACTIVITIES

1. Go to the library and find who discovered X rays and when they were discovered. List some medical applications of X rays.
2. Go to the library and research magnetic fields. What are their potential hazards? Has it been proven that magnetic fields are dangerous to humans?

REFERENCES

Bar-Cohen, Y., and A.K. Mal. *Ultrasonic Testing, Vol. 8, ASNT Handbook*. American Society for Nondestructive Testing.

Dobmann, G., 1985. Magnetic Leakage Flux Techniques in NDT: A State of the Art Survey of the Capabilities for Defect Detection and Sizing. *Electromagnetic Methods of NDT*, ed. W. Lord, Gordon and Breach.

Forster, F., 1986. Magnetic Findings I: The Fields of Nondestructive Magnetic Leakage Field Inspection. *NDT International*, Vol. 19.

General Dynamics. 1977. *Nondestructive Testing Magnetic Particle*. Classroom Training Handbook, 2nd ed.

General Dynamics. 1983. *Nondestructive Testing: Radiographic Testing*. Classroom Training Handbook.

General Dynamics. *Nondestructive Testing: Eddy Current Testing*. Classroom Training Handbook.

Halmshaw, R., 1982. Industrial Radiology: Theory and Practice. *Applied Science*.

Industrial Radiography. 1986. Mortsel, Agfa-Gevaert Handbook, revised ed.

Knowll, G.F., 1979. *Radiation Detection and Measurement.* John Wiley & Sons.

Krautkramer, H., and J. Krautkramer, 1983. *Ultrasonic Testing of Materials*, 3rd ed. Springer-Verlag.

Libby, H.L., 1971. *Introduction to Electromagnetic Nondestructive Test Methods.* John Wiley & Sons.

Morgan, K.Z., and J.E. Turner, 1973. *Principles of Radiation Protection.* John Wiley & Sons.

Radiography and Radiation Testing. 1985. *Nondestructive Testing Handbook, Vol. 3,* 2nd ed. American Society for Nondestructive Testing.

Metals Handbook 1989. *Nondestructive Evaluation and Quality Control. Vol. 17,* 9th ed.

Appendix

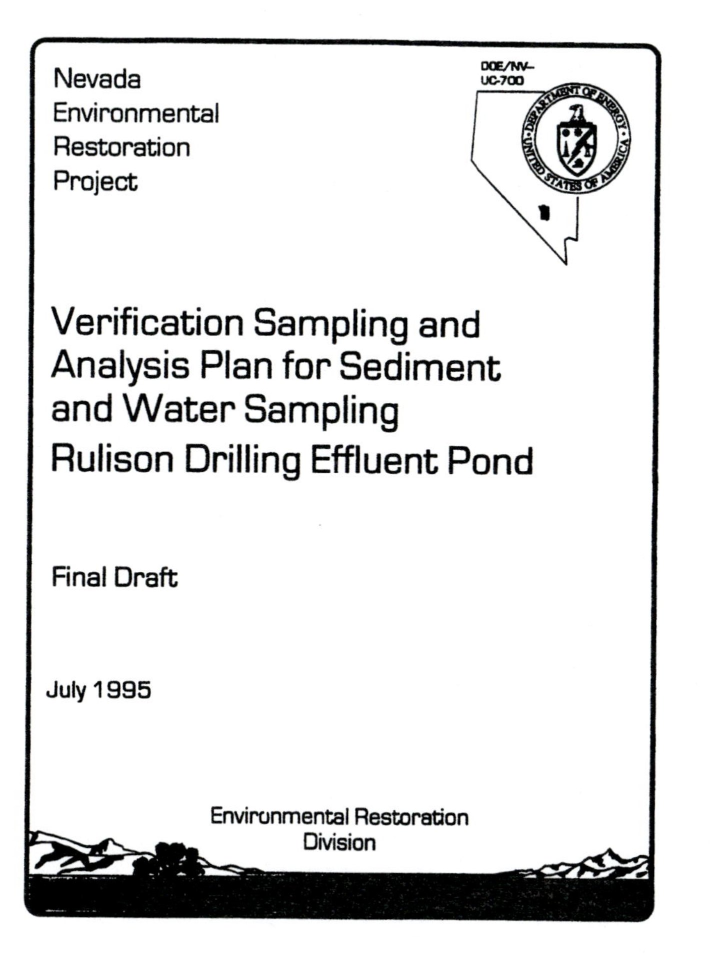

Nevada
Environmental
Restoration
Project
DOE/NV-
UC-700
DEPARTMENT OF ENERGY
UNITED STATES OF AMERICA
Verification Sampling and
Analysis Plan for Sediment
and Water Sampling
Rulison Drilling Effluent Pond
Final Draft
July 1995
Environmental Restoration
Division

Table of Contents

Table of Contents (Continued)

Table of Contents (Continued)

List of Tables

List of Acronyms and Abbreviations

AR/COC	Analysis Request and Chain of Custody
BTEX	benzene, toluene, ethylbenzene, xylene
CAP	Corrective Action Plan
COC	contaminant of concern
DOE	U.S. Department of Energy
DOE/NV	U.S. Department of Energy, Nevada Operations Office
DQO	Data Quality Objective
EPA	U.S. Environmental Protection Agency
ESSC	Environmental Services Support Contractor
FAC	Field Activities Coordinator
ft	foot
m	meter
mℓ	milliliter
MS/MSD	Matrix Spike and Matrix Spike Duplicate
QAPP	Quality Assurance Project Plan
QC	quality control
RCRA	Resource Conservation and Recovery Act
SSHASP	Site-Specific Health and Safety Plan
TCLP	Toxicity Characteristic Leaching Procedure
TDS	total dissolved solid
TPH	total petroleum hydrocarbon
VSAP	Verification Sampling and Analysis Plan
VOA	volatile organic analysis
yd^3	cubic yard
°C	degree Celsius

1.0 Introduction

The purpose of this Verification Sampling and Analysis Plan (VSAP) is to provide guidance for collecting and analyzing soil, water, and sediment samples during the remediation of the Rulison Site drilling effluent pond. This plan provides guidance for activities associated with the collection of

- water samples prior to, during, and following pond drainage and construction dewatering
- stabilized sediment samples prior to shipment to a disposal landfill
- water samples from the pond water which must be treated prior to discharge
- soil samples from under the former pond sediments to verify clean closure of the pond.

1.1 Project Description and Background

Project Rulison was a joint U.S. Atomic Energy Commission and Austral Oil Company experiment conducted as part of the Plowshare Program. The experiment was conducted to test the feasibility of using a nuclear device to increase natural gas production in low-permeability, gas-producing geologic formations. The experiment was conducted on September 10, 1969, and consisted of detonating a 43-kiloton device at a depth of 2,568 meters (m) (8,426 feet [ft]) below the ground surface. Production testing of the well was conducted in 1970 and 1971. The site was cleaned up in 1972 with a final cleanup conducted after the testing wells were plugged in 1976. Some surface contamination resulted from decontamination of drilling equipment and fallout from the gas flaring (DOE, 1988). Except for cleanup of the drilling effluent pond, all surface contamination was removed during site clean-up operations.

The site is situated on the north slope of Battlement Mesa on the upper reaches of Battlement Creek, at an elevation of approximately 2,500 m (8,200 ft). The valley is open to the north-northwest and is bounded on the other three sides by steep mountain slopes that rise to elevations above 2,927 m (9,600 ft). The drilling effluent pond is triangular in shape and covers approximately 0.5 acre. It is approximately 6 m (20 ft) deep from the top of the berm to the pond bottom and is located approximately 400 m (1,312 ft) north-northwest of the original surface ground zero well. The pond originally contained drilling fluids, but was converted to a fresh-water trout pond. The pond was left in place at the request of the land owner (ERDA, 1977) and contains aquatic vegetation, tiger salamanders *(Ambystoma tigrinium)*, and stocked rainbow trout. The pond is fenced to prevent access to wildlife and livestock.

1.2 Previous Studies

The drilling effluent pond at the Rulison Site was used to store nonradioactive drilling wastes resulting from drilling of the device emplacement hole (Well R-E). Cuttings and most of the drilling fluid were excavated, transported off site, and properly disposed in 1976; however, some residual fluid was left in the pond. The drilling fluids consisted of a bentonitic drilling mud with various additives used to improve drilling characteristics, such as diesel fuel and chrome lignosulfonate. In 1994 and 1995, three pond-sediment sampling events were conducted to evaluate the nature of this residual drilling fluid. Surface water, soil, and sediment samples were collected. All analytical results of surface water samples were clean, with no petroleum compounds or metals present. The results of the sampling events are presented in the Rulison Site Corrective Action Plan (CAP) (DOE, 1995a) and in Tables 5-1 through 5-6. Also included are state of Colorado and federal clean-up standards for heavy metals and organic compounds in soils. Colorado discharge standards for water are site-specific and will be specified in the water discharge permit.

1.3 Sampling Objectives and Approach

There are four objectives of this sampling and analysis event. First, samples of pond water will be collected before, during, and following pond drainage activity to verify that the discharge water is not contaminated. Also, samples of water produced during construction dewatering activities, if necessary, will be also sampled to verify that discharged water was not contaminated. Second, samples of stabilized sediment will be collected to verify that the sediment complies with the landfill waste acceptance criteria following stabilization. Third, it is anticipated that as water drainage from the pond nears completion, the water will become mixed with contaminated sediment and will have to be stored on site in frac (Baker) tanks, treated, and then discharged. This water will be sampled prior to discharge to verify that treatment process effectiveness meets all federal and state discharge standards. Finally, representative samples of soil from underneath the drilling effluent pond will be collected to verify that all contaminated sediment from the bottom of the pond was removed during pond remediation and that constituents of concern from the pond did not contaminate the underlying soil.

The proposed approach for sampling of water drained from the pond and discharged during construction dewatering will be as follows:

- Water will be drained from the pond into the nearby stream until the water level in the pond has been reduced as much as possible as specified in the CAP (DOE, 1995a).

- If construction dewatering is necessary, this produced water will also be discharged to the same nearby stream.

- Water samples will be collected at frequencies as specified in Chapter 5.0 of this VSAP, utilizing approved contractor procedures and will be analyzed according to methods specified in the Rulison Site Quality Assurance Project Plan (QAPP) (DOE, 1995b).

- Water samples will be collected prior to and in the initial stages of drainage to provide the baseline for which subsequent samples will be compared.

- Additional water samples will be collected at the midpoint of pond drainage and again near the end of pond drainage.

- Water samples will be analyzed for whole effluent toxicity; total Resource Conservation and Recovery Act (RCRA) metals; total petroleum hydrocarbons (TPH) as diesel; and benzene, toluene, ethylbenzene, and total xylenes (BTEX).

The proposed approach for sampling of sediment following stabilization and prior to shipment to the designated landfill will be as follows:

- A large, truck-mounted pug mill will be used for mixing the sediment with the proprietary stabilizer, as specified in the Rulison CAP (DOE, 1995a).

- Samples of stabilized sediment will be collected at regular intervals during the stabilization process using approved contractor procedures.

- Stabilized sediment samples will be analyzed for Toxicity Characteristics Leaching Procedure (TCLP) chromium, TPH (diesel), and TCLP Benzene, according to methods defined in the Rulison Site QAPP (DOE, 1995b).

- Specific information regarding sampling and analysis stabilized sediments is provided in Section 5.0.

The proposed approach for sampling water that may have become mixed with contaminated sediment following removal from the pond will be as follows:

- All potentially contaminated water will be stored on site during the final stages of removal from the pond in frac (Baker) tanks.

- This water will be treated and discharged directly if it passes the discharge criteria. The water will not be discharged until it meets state and federal water quality standards.

- Samples of the treated water will be collected using approved contractor procedures.

- Treated water samples will be analyzed for total RCRA metals, TPH (diesel), and BTEX, according to methods defined in the Rulison Site QAPP (DOE, 1995b).
- Specific information regarding sampling and analysis of potentially contaminated water is provided in Section 5.0.

The proposed approach for collecting samples of soil from underneath the former pond will be as follows:

- During or following the removal of all remaining water and sediment from the pond (depending on site conditions), the ground surface will be sampled at frequencies specified in Chapter 5.0 of this VSAP to verify that all contaminated sediment from the bottom of the pond was removed during pond reclamation and that constituents of concern from the pond did not contaminate underlying soil.
- The frequency and locations of soil samples will be selected according to guidance provided by *Methods for Evaluating the Attainment of Cleanup Standards, Volume 1: Soils and Solid Media* (EPA, 1989a); *Verification of PCB Spill Cleanup By Sampling and Analysis* (EPA, 1985); *Statistical Analysis of Ground-Water Monitoring Data at RCRA Facilities, Interim Final Guidance* (EPA, 1989b); and *Statistical Analysis of Ground-Water Monitoring Data at RCRA Facilities, Addendum to Interim Final Guidance* (EPA, 1992).
- The soil samples will be collected using approved contractor procedures.
- Soil samples will be analyzed for total TCLP RCRA metals, TPH (diesel), and BTEX, according to methods defined in the Rulison Site QAPP (DOE, 1995b).
- If the analytical results of a particular sample are above the regulatory limit, additional soil shall be removed from that sampling location until the state of Colorado soil clean-up standard is met.

1.4 Project Schedule

The proposed target starting date for field activities is July 1995. The field activities are expected to last 90 working days. Pending review and approval of this VSAP, sampling activities will be definitively scheduled. The target date for completion of this project is September 30, 1995.

1.5 Document Ownership

The U.S. Department of Energy (DOE), Nevada Operations Office (DOE/NV), Nevada Environmental Restoration Project, is the owner of this VSAP. This department is responsible for initiating the document review process; the Environmental Services Support Contractor (ESSC) is responsible for incorporating or resolving reviewer comments and concerns into the final plan. Questions or comments concerning this document should be addressed to the DOE/NV Rulison Site Manager and/or the ESSC Project Manager.

2.0 *Project Organization and Responsibilities*

The ESSC is tasked with managing the verification sampling and analysis activities. The duties of organizations and individuals participating in the off-site background sample collection effort are outlined below.

2.1 Duties

The DOE Rulison Site Manager has the following duties:

- Requests and allocates resources for Rulison Site activities
- Coordinates Rulison Site activities
- Ensures that sampling and analytical activities be conducted in accordance with DOE guidelines, as well as with other applicable state and federal regulations
- Responds to recommendations from audits and assessments of the Rulison Site
- Reviews and approves plans necessary to control the quality of sampling and analytical data.

The ESSC Project Manager has the following duties:

- Arranges for preparation, review, and approval of the VSAP; distributes the approved VSAP; and revises the VSAP, as needed
- Arranges for an Industrial Hygiene representative to review proposed sampling activities and a Site-Specific Heath and Safety Plan (SSHASP)
- Provides for the analytical cost estimation and budget preparation for sampling and analytical activities
- Assigns a qualified analytical laboratory
- Tracks analytical invoice and processing coordination for payment
- Interfaces and resolves problems between the field and the laboratory
- Conducts quality control (QC) check of all field documentation
- Notifies all relevant personnel of sampling schedules and ensures that entry into the site is arranged
- Oversees the implementation of this VSAP

- Evaluates project changes, nonconformances, and corrective actions and notifies affected personnel

- Ensures that original copies of all field forms, chain of custody and request for analysis records, analytical data, sample collection, and equipment maintenance logs are entered into document control

- Assesses analytical data generated under this VSAP.

Sampling will be conducted by a field crew consisting of a Field Activities Coordinator (FAC) and sampling personnel. The FAC is the on-site representative and is responsible for the following:

- Scheduling analytical services
- Procuring sampling equipment and containers
- Developing, maintaining, and implementing the approved SSHASP
- Providing day-to-day management of the sampling team
- Supervising sample collection
- Reviewing all field documentation
- Packaging, transporting, and shipping samples to the laboratory
- Tracking sample information (for sample number, chain of custody number, request for analysis number, sample type [e.g., soil and water], contract laboratory, shipping date, sample location, project contact, and priority of sample)
- Monitoring QC analyses performed by the laboratory
- Verifying laboratory analytical reports
- Overseeing field site control
- Ensuring that sampling personnel have proper documentation of appropriate Health and Safety training while in the field
- Maintaining field notes
- Reporting nonconformances and perform corrective actions.

The ESSC Health and Safety Officer is responsible for the following:

- Reviewing the SSHASP
- Providing field activities oversight, as needed.

The Analytical Laboratory is responsible for the following:

- Preparing and analyzing samples
- Maintaining chain of custody documentation
- Validating initial data
- Reporting nonconformances and perform corrective actions
- Reporting data
- Submitting data summary packages that meet data quality requirements.

2.2 Personnel Training and Qualifications

Prior to conducting sample collection activities, field personnel must complete health and safety training for hazardous waste workers that conforms to U.S. Occupational, Safety, and Health Administration requirements found in Title 29 Code of Federal Regulations §1910.120(E). Sampling team members must demonstrate knowledge of sampling procedures and equipment operation, in accordance with the Rulison Site QAPP, gained through documented formal training and/or on-the-job experience. Health-related qualifications include initial and periodic physical examinations.

Prerequisite general hazards training for soil sampling shall include instruction in employee right-to-know issues, chemical and physical hazards associated with sampling, and safe work practices. General hazards training shall be repeated annually for each employee. No other specialized safety training is required for persons working under this VSAP.

3.0 Data Quality Objectives

Data Quality Objectives (DQO) are qualitative and quantitative statements derived from the outputs of the DQO planning process. The DQO process is a series of planning steps designed to ensure that the type, quantity, and quality of environmental data used in decision making is appropriate for the intended application (EPA, 1993). The DQO process, for the purposes of this project, can be divided into four major elements: a statement of sample collection objective(s), DQO development, an analytical data assessment, and the development of a VSAP that satisfies the DQOs.

Data Quality Objectives have been established to define the data quality requirements necessary to meet the project objectives stated in Section 1.1.2 of this VSAP. Chemical analysis of soils will meet the DQO guidance criteria presented in *Data Quality Objectives for Remedial Response Activities, Development Response* (EPA, 1987a).

3.1 Sample Collection Objectives

The main objectives of this sample collection project are presented in Section 1.1.2 of this VSAP.

3.2 Analytical Data Assessment

Chemical data used for this project will be consistent with the U.S. Environmental Protection Agency (EPA) data verification and validation procedures outlined in SW-846 (EPA, 1987b). Sample analysis data will be assessed using the following indicators:

- Laboratory quantitation limits
- Laboratory analysis bias
- Laboratory analysis precision
- Data completeness
- Sample representativeness
- Data comparability.

3.3 Laboratory Quantitation Limits

Chapter 4.0 provides laboratory quantitation limits expected for this task. Failure to attain these limits may result in the conditional acceptance or rejection of analytical data.

3.4 Laboratory Analysis Bias

Bias acceptance criteria will be reported by the laboratory in the analytical data report for each analyte allowing for evaluation of the control sample results. Control sample acceptance criteria are defined as plus or minus three standard deviations from the mean percent recovery of at least 20 laboratory control samples. Laboratory prepared method blanks are used to monitor bias from contamination introduced during analytical procedures. Positive values for method blanks can qualify analytical results in the associated investigatory samples indicating false positive results.

3.5 Laboratory Analysis Precision

Precision is assessed by means of duplicate or replicate sample analysis. For this project, precision will be measured through analysis of duplicate laboratory control samples. Precision is monitored by the laboratory in the same way as bias. Precision acceptance criteria will be included in the analytical data report for evaluation of analytical precision.

3.6 Data Completeness

The data completeness goal for this project is 80 percent because of the limited number of samples from a given location.

3.7 Sample Representativeness

Duplicate samples will be collected to document sampling representativeness. If analyses results for the sample and its duplicate do not differ substantially, the sampling method will be determined sufficient and the samples representative.

4.0 Quality Control

Quality control for sampling at all locations will be implemented to ensure that the measurement data collected meet the DQOs for this investigation. Quality control will be implemented by strict adherence to the sampling procedures described in Chapters 5.0 and 6.0 of this plan; documenting the sampling activities and sample custody; using standard equipment and materials; and collecting, analyzing, and evaluating field and laboratory QC samples. Field and laboratory QC shall be maintained and documented in accordance with the Rulison Site QAPP. Specific activities are outlined below.

4.1 Field Quality Control

Sample collection will be performed in strict accordance with this VSAP and approved contractor procedures. Samples will be collected in properly cleaned, laboratory-prepared containers, using equipment that has been properly decontaminated. Field QC samples will be collected as indicated below.

4.2 Field Duplicate Samples

Duplicate environmental samples will be collected at a rate of 10 percent of the original samples and analyzed for the same suite of analytes to assess subsample variability. Environmental duplicates will not be taken from the same locations as the matrix spike and matrix spike duplicate (MS/MSD) samples.

4.3 Matrix Spike and Matrix Spike Duplicate Samples

Matrix spike and matrix spike duplicate samples will be requested at a rate of 5 percent of the number of environmental samples and analyzed to determine interferences of the sample matrices on the analytical methods and subsample variance of the composite samples. The MS/MSD samples will not be specified from the same location as the duplicate samples. The MS/MSD aliquots shall be taken from each environmental sample designated by the field supervisor on the Analysis Request and Chain of Custody (AR/COC) Record.

4.4 Equipment Rinsate Blank Samples

Equipment rinsate blanks will be collected at a rate of 10 percent of the number of environmental samples and analyzed for the same suite of analytes as the samples. Equipment rinsate samples will be collected immediately following final decontamination of the sampling device.

4.5 Trip Blank Samples

Trip blank samples will be used during the project to document the occurrence of contamination of samples during transport to the analytical laboratory. The trip blanks will be prepared by the laboratory and shipped to the site with the sample containers. Trip blanks will consist of two 40-milliliter (mℓ) glass volatile organic analysis (VOA) vials filled with deionized water at the laboratory. One set of trip blanks will accompany each shipping cooler with BTEX samples. These samples will be subjected to the same sample management and documentation procedures as the environmental BTEX samples.

4.6 Field Blank Samples

One deionized-water field blank will be collected from each source of deionized water used during the project to verify the decontamination water chemistry. The sample will be collected by pouring deionized water directly into the appropriate sample bottles. These samples will be subjected to the same sample management and documentation procedures as the environmental samples.

4.7 Laboratory Quality Control Samples

Laboratory QC will be maintained in accordance with the Rulison Site OAPP using the standard procedures established by the laboratory. Method blank, laboratory control sample, and laboratory-control-sample duplicate samples will be analyzed and used to evaluate method and instrumental accuracy and precision.

4.8 Calculation of Data Quality Indicators

Analytical data quality will be assessed in part using the indicators for bias, precision, and completeness in accordance with procedures specified in the Rulison Site QAPP.

5.0 Sampling Strategy and Sampling Locations

Sampling activities will be conducted according to the sampling strategy, methodology, and sampling locations selected as detailed below. Samples will be collected using approved contractor procedures. All sample collection activities shall be thoroughly documented on the Sample Collection Log for each sample. Daily activities shall be recorded sequentially on a Field Activity Daily Log.

Following verification sampling activities, the remainder of site restoration activities will be completed as part of the Rulison CAP (DOE, 1995a).

5.1 Selection of Sampling Frequency and Sampling Locations

The rationale and approaches outlined below were used for determining optimal sampling frequency and locations for each of the sampling tasks of this project.

5.2 Water Samples prior to, during, and following Pond Drainage, and during Construction Dewatering

The goal of this sampling task is to verify that the water being drained into the nearby stream is not contaminated from the sediment remaining in the bottom of the pond or that water discharged as part of construction dewatering activities is not contaminated. Sample analytical results will be compared to the state of Colorado water clean-up standards specified in the discharge permit to verify that discharge criteria have been met. The following approach will be used:

- One sample will be collected from the pond prior to initiating discharge and will be analyzed for Whole Effluent Toxicity.

- One sample will be collected at the beginning of pond drainage activity, and analyzed for total lead, chromium, and barium, according to the methods cited in the Rulison Site QAPP for this project (DOE, 1995b).

- At the midpoint of draining the pond, one sample will be collected for suspended solids; total dissolved solids (TDS); TPH (diesel); and BTEX. Samples will be analyzed according to methods defined in the Rulison Site QAPP.

- At the end of pond drainage activity, one sample will be collected for TDS, TPH (diesel), and BTEX. Samples will be analyzed according to methods defined in the Rulison Site QAPP.

In addition, discharged water samples shall be collected and analyzed during pond clean-up operations at the discretion of the site supervisor in the event that site conditions change or additional water-quality information is required. Appropriate QC samples shall be collected along with these water samples in accordance with the Rulison Site QAPP.

5.3 Stabilized Sediment Samples

The goal of this sampling task is to verify the concentrations of constituents of concern in the sediment following stabilization for the purposes of determining whether the stabilized sediment represents a hazardous waste. Analytical results will be provided to the Colorado Department of Health and the Environment and to a specified landfill owner/operator, for a hazardous or nonhazardous determination prior to anticipated disposal at the landfill. Approximately 3,000 cubic yards (yd^3) of stabilized sediment will be produced, depending on the final method of stabilization. Samples of stabilized sediment will be collected as the sediment is prepared for shipment to the specified landfill for disposal. Ten samples of the stabilized sediment will be collected, corresponding to an approximate rate of 1 sample for every 300 yd^3 of stabilized sediment. Samples will be collected according to approved contractor procedures. Samples will be analyzed for TCLP chromium, TPH (diesel), and TCLP Benzene according to methods specified in the Rulison Site QAPP (DOE, 1995b), for the purposes of comparison with the unmodified sediment sample results.

5.4 Treated Pond Water Samples

The goal of this sampling task is to verify that water, drained from the pond, but contained on site in frac (Baker) tanks, meets federal and state discharge criteria following treatment. The treatment methodologies are specified in the Rulison CAP (DOE, 1995a). Following treatment, two water samples will be collected from each container of treated water to test for discharge criteria. The samples will be collected using approved contractor procedures. The samples will be analyzed for total RCRA metals, TPH (diesel), and BTEX according to methods specified in the Rulison Site QAPP (DOE, 1995b). Sample analytical results will be compared to state of Colorado water clean-up standards specified in the discharge permit to verify that discharge criteria have been met.

5.5 Verification Soil Samples

The goal of this sampling task is to acquire representative samples of soil from underneath and adjacent to the drilling effluent pond to verify that all contaminated sediment from the bottom of the pond was removed during pond remediation and that constituents of concern from the pond did not contaminate underlying soil. A statistical approach based on EPA guidance is used; however, due to the lack of information about the nature and extent of contaminants of concern

(COC) in soils below the pond sediments, the number of proposed soil samples estimated below is conservative and may be modified based on actual site conditions.

The frequency and locations of soil samples will be selected according to guidance provided primarily by *Methods for Evaluating the Attainment of Cleanup Standards, Volume 1: Soils and Solid Media* (EPA, 1989a), and by *Verification of PCB Spill Cleanup By Sampling and Analysis* (EPA, 1985). Additional statistical guidance will be obtained from *Statistical Analysis of Ground-Water Monitoring Data at RCRA Facilities, Interim Final Guidance* (EPA, 1989b), and *Statistical Analysis of Ground-Water Monitoring Data at RCRA Facilities, Addendum to Interim Final Guidance* (EPA, 1992).

A systematic sampling approach will be used to assess the occurrence of any soil contamination. This approach is chosen to distribute the samples more uniformly over the site. Because the sample points follow a simple pattern and are separated by a fixed distance, locating the samples points in the field will be easier than if a random approach was selected. This method will also minimize the possibility that more contaminated areas of the site will not be represented in the sample (EPA, 1989c). The soil samples will be collected using approved contractor procedures. All soil samples will be analyzed for TPH (diesel) and BTEX. In addition, samples will be analyzed for TCLP RCRA metals at the rate of 20 percent of the TPH and BTEX samples. Samples will be analyzed according to methods defined in the Rulison Site QAPP (DOE, 1995b). Sample analytical results will be compared to state of Colorado soil clean-up standards to verify that clean-up criteria have been met. If the analytical results of a particular sample are above the regulatory limit, then additional soil shall be removed from that sampling location until the state of Colorado soil clean-up standard is met. The FAC may require additional or fewer samples depending on actual or changing site conditions.

Analytical results for samples collected to date from the site are presented in Tables 5-1 through 5-6. As may be seen from the data, primary constituents of concern are TPH (diesel), chromium (total), and, to a lesser extent, lead, barium, and BTEX components. Statistical analysis of the data indicate that rigorous calculations of sample population size needed to establish clean-up attainment are not possible for all COCs, particularly chromium (total) and TPH (diesel), due to the high degree of variation and nonnormal distribution of the analytical results.

Table 5 1
Sediment Sampling Results for Rulison Drilling Effluent Pond
(Page 1 of 3)

Compound	Regulatory Limit	Sampling Conducted in September 1994											
		Sediment Samples											
		SD-01	Q[a]	SD-02	Q	SD-03	Q	SD-04	Q	SD-05	Q	EQ-01[c]	Q
Total Metals (mg/kg)												µg/ℓ	
Aluminum		11,700		13,400		32,300		61,500		30,700		56.7	B
Antimony		0.72	B[a]	0.57	U[a]	0.81	U	1.5	U	3.7	U	1.8	U
Arsenic	100[b]	7.6		7.1		15.5		31.6		12.9	B	1.8	U
Barium	2,000[b]	158		79		395		1,140		816		2.2	B
Beryllium		0.79	B	0.78	B	1.8	B	4.8		2.4	B	0.39	B
Cadmium	20[b]	1.3	U	1.2	U	1.7	U	3.1	U	7.6	U	3.7	U
Calcium		18,800		17,800		16,700		53,500		37,300		26.5	B
Chromium	100[b]	20.6		29.7		55.9		**114**		**2,170**		3.4	U
Cobalt		8.3	B	8.9	B	15.8	B	34.4	B	19.1	B	3.2	U
Copper		20.5		22.1		47.7		95.8		164		7.4	B
Iron		17,900		16,100		36,300		71,300		37,200		66.1	B
Lead	100[b]	13.2		8.5		30.6		68.9		**427**		1.4	B
Magnesium		7,360		6,540		12,800		29,500		16,900		84.7	B
Manganese		243		287		670		1,460		883		2.1	B
Mercury	4[b]	0.11	B	0.08	B	0.11	U	0.42		0.90	B	0.10	U
Nickel		17.2		20.4		42.1		89.3		60.7	B	15.5	U
Potassium		2,200		1,990		3,890		12,500		8,620	B	1,940	U
Selenium	20[b]	0.50	B	0.41	U	0.59	U	1.1	U	2.7	U	1.3	U
Silver	100[b]	1.3	U	1.2	U	1.7	U	3.1	U	7.6	U	3.7	U
Sodium		820	B	505	B	852	B	5,220		1,970	B	459	B
Thallium		0.51	U	0.47	U	0.68	U	1.2	U	3.1	U	1.5	U
Vanadium		38.1		35.3		75.5		129		57.1	B	3.1	U
Zinc		58.3		49.5		103		178		191		12.5	B

Refer to footnotes at end of table.

Table 5-1
Sediment Sampling Results for Rulison Drilling Effluent Pond
(Page 2 of 3)

Compound	Regulatory Limit	SD-01	Q[a]	SD-02	Q	SD-03	Q	SD-04	Q	SD-05	Q	EQ-01[c]	Q
	Sampling Conducted in September 1994												
		Sediment Samples											
TCLP Metals (mg/ℓ)												mg/ℓ	
Chromium	5[d]	NA[e]		NA		NA		NA		0.066		NA	
Lead	5[d]	NA		NA		NA		NA		0.042	U	NA	
TPH (mg/kg)												mg/ℓ	
Nonspecific	250[f]	NA		15.8	U	NA		17,000		72,600		NA	
Gas		NA		NA		NA		NA		NA		NA	
Diesel		NA		NA		NA		NA		NA		NA	
Waste Oil		NA		NA		NA		NA		NA		NA	
BTEX (μg/kg)												μg/ℓ	
Benzene	g	NA		NA		NA		NA		NA		NA	
Toluene	g	NA		NA		NA		NA		NA		NA	
Ethylbenzene	g	NA		NA		NA		NA		NA		NA	
Xylene	g	NA		NA		NA		NA		NA		NA	
Total BTEX	50,000	NA		NA		NA		NA		NA		NA	
PCB (μg/kg)												μg/ℓ	
Aroclor-1016		NA		NA		NA		140	U	340	U	NA	
Aroclor-1221		NA		NA		NA		140	U	340	U	NA	
Aroclor-1232		NA		NA		NA		140	U	340	U	NA	
Aroclor-1242		NA		NA		NA		140	U	340	U	NA	
Aroclor-1248		NA		NA		NA		140	U	340	U	NA	
Aroclor-1254		NA		NA		NA		140	U	340	U	NA	
Aroclor-1260		NA		NA		NA		140	U	340	U	NA	

Refer to footnotes at end of table.

Table 5-1
Sediment Sampling Results for Rulison Drilling Effluent Pond
(Page 3 of 3)

Compound	Regulatory Limit	Sampling Conducted in September 1994											
		Sediment Samples											
		SD-01	Q[a]	SD-02	Q	SD-03	Q	SD-04	Q	SD-05	Q	EQ-01[c]	Q
Gross Alpha/Beta (pCi/g)												pCi/ℓ	
Gross Alpha		15.6		14.6		7.72		11.6		6.56		0.27	
2 Sigma Error (+/-)		5.0		4.9		3.74		4.5		3.58		0.16	
MDA[h]		5.0		5.0		5.18		5.3		5.23		0.21	
Gross Beta		25.8		24.4		22.4		20.6		17.4		-0.003	
2 Sigma Error (+/-)		3.9		3.7		3.4		3.3		2.9		0.046	
MDA		3.7		3.6		3.2		3.5		3.1		0.665	
Gamma Spec. (pCi/g)												pCi/ℓ	
Cesium-137		ND[i]		ND		ND		ND		ND		ND	
Potassium-40		22.1		24.4		17.5		15.2		11.2		ND	
Lead-212		ND		1.30		ND		1.06		1.23		ND	
Radium-226		0.91		0.75		ND		ND		ND		ND	

[a] Q = Laboratory assigned data qualifier: U = Compound was analyzed for but not detected; 3 = In organics, the analyte was found in the blank. In inorganics, the result is above the Instrument Detection Limit but below the Contract Required Detection Limit.
[b] No regulations for these soil parameters are specified in the Colorado Department of Healt ı "Storage Tank Facility Owner/Operator Guidance Document". Regulatory limits are based on 20X the RCRA "Maximum Concentration of Contaminants for the Toxicity Characteristic".
[c] Quality Assurance Sample
[d] No regulations for these soil parameters are specified in the Colorado Department of Healt ı "Storage Tank Facility Owner/Operator Guidance Document". Regulatory limits are based on RCRA "Maximum Concentration of Contaminants for the Toxicity Characteristic".
[e] The sample was not analyzed for that parameter.
[f] This limit is based on regulations specified in the Colorado Department of Health "Storage Tank Facility Owner/Operator Guidance Document".
[g] No individual regulatory level for this parameter, it is combined as Total BTEX.
[h] Minimum Detectable Activity
[i] Nondetect means the analyte was not found in the sample at a concentration above the instrument detection limit.

mg/kg = milligram per kilogram
pCi/g = picocurie per gram
pCi/ℓ = picocurie per liter
mg/ℓ = milligram per liter
μg/ℓ = microgram per liter
μg/kg = microgram per kilogram

Table 5-2
Sediment Sampling Results for Rulison Drilling Effluent Pond
(Page 1 of 2)

Compound	Regulatory Limit	Sampling Conducted in October 1994																	
		Sediment Samples																	
		SD-06	Q[a]	SD-07	Q	SD-08	Q	SD-09	Q	SD-10	Q	SD-11	Q	SD-12	Q	SD-13[b]	Q	WFR-O[c]	Q
Total Metals (mg/kg)																		μg/l	
Aluminum		7,830		1,930		2,300		3,270		3,250		4,160		1,830		2,160		37.4	B
Antimony		0.36	U	0.36	U	0.36	U	0.36	U	0.36	U	0.36	U	0.36	U	0.36	U	1.8	U
Arsenic	100[d]	3.5		0.60	B	0.97	B	0.72	B	0.56	B	2.3		1.1	B	0.69	B	1.0	U
Barium	2,000[d]	132		100		87.8		152		174		96.3		71.3		88.0		1.8	B
Beryllium		0.56	B	0.39	B	0.39	B	0.39	B	0.43	B	0.39	B	0.26	B	0.30	B	0.29	B
Cadmium	20[d]	1.1		0.74	U	0.74	U	0.74	U	0.74	U	0.74	U	0.74	U	0.74	U	3.7	U
Calcium		16,100		5,340		5,960		7,060		6,690		13,600		6,510		5,960		130	B
Chromium	100[d]	22.3		187		233		343		317		106		214		206		3.5	B
Cobalt		5.0	B	1.4	B	1.6	B	2.0	B	2.6	B	1.8	B	0.84	B	1.5	B	3.2	U
Copper		17.0		9.0		10.9		13.4		9.6		11.6		7.3		10.6		10.0	B
Iron		11,700		3,410		3,980		5,560		5,000		5,380		2,670		3,570		72.2	B
Lead	100[d]	8.3		10.1		11.3		13.3		9.2		8.1		28.8		13.9		1.0	U
Magnesium		4,300		1,590		1,930		2,220		2,230		2,590		1,250		1,760		133	B
Manganese		148		79.8		115		111		99.2		81.6		74.3		106		3.6	B
Mercury	4[d]	0.05	U	0.05	U	0.05	U	0.05	U	0.05	U	0.05	U	0.05	U	0.05	U	0.16	B
Nickel		15.6		4.3	B	4.8	B	7.3	B	8.[illegible]	B	5.1	B	3.5	B	5.4	B	15.5	U
Potassium		1,560		527	B	902	B	1,350		1,420		877	B	389	U	553	B	1,940	U
Selenium	20[d]	0.28	B	0.26	U	0.26	U	0.26	U	0.26	U	0.26	U	0.26	U	0.26	U	1.3	U
Silver	100[d]	0.74	U	0.74	U	0.74	U	0.74	U	0.74	U	0.74	U	0.74	U	0.74	U	3.7	U
Sodium		368	B	264	B	218	B	1,630		973	B	351	B	233	B	288	B	181	B
Thallium		0.30	U	0.30	U	0.30	U	0.30	U	0.30	U	0.30	U	0.30	U	0.30	U	1.5	U
Vanadium		19.2		3.8	B	4.0	B	6.1	B	6.8	B	10.9		3.7	B	4.1	B	4.9	B
Zinc		36.7		21.3		22.7		29.5		23.6		23.0		14.0		20.1		7.4	B

Refer to footnotes at end of table.

Table 5-2
Sampling Results for Rulison Drilling Effluent Pond
(Page 2 of 2)

Compound	Regulatory Limit	Sampling Conducted in October 1994																	
		Sediment Samples																	
		SD-06	Q[a]	SD-07	Q	SD-08	Q	SD-09	Q	SD-10	Q	SD-11	Q	SD-12	Q	SD-13[b]	Q	WFR-O[c]	Q
TCLP Metals (mg/ℓ)																			
Chromium	5[e]	NA[f]		NA		NA		0.44		NA		NA		NA		NA		NA	
Lead	5[e]	NA		NA		NA		NA		NA		NA		NA		NA		NA	
TPH (mg/kg)																		µg/ℓ	
Nonspecific	250[g]	NA		NA		NA		NA		NA		NA		NA		NA		NA	
Gas		0.50	U	250		28		79		260		260		210		7.6	*	100	U
Diesel		24	U	4,800		15,000		9,600		11,000		4,400		10,000		12,000		500	U
Waste Oil		34		2,500	U	490	U	250	U	2,400	U	250	U	240	U	500	U	500	U
BTEX (µg/kg)																		µg/ℓ	
Benzene	h	2.0	U	27		2.0	U	26		14		26		19		2.0	U	NA	
Toluene	h	2.0	U	690		9.5		660		700		310		370		31		NA	
Ethylbenzene	h	2.0	U	980		29		880		990		890		1,200		62		NA	
Xylene	h	2.0	U	4,300		160		3,800		4,400		4,100		5,200		300		NA	
Total BTEX	50,000		U	5,997		200.5		5,366		6,104		5,326		6,789		395		NA	

[a] Q = Laboratory assigned data qualifier: U = Compound was analyzed for but not detected; B = In organics, the analyte was found in the blank. In inorganics, the result is above the Instrument Detection Limit but below the Contract Required Detection Limit.
[b] Duplicate sample of SD-08
[c] Quality Assurance Sample
[d] No regulations for these soil parameters are specified in the Colorado Department of Health "Storage Tank Facility Owner/Operator Guidance Document". Regulatory limits are based on 20X the RCRA "Maximum Concentration of Contaminants for the Toxicity Characteristic".
[e] No regulations for these soil parameters are specified in the Colorado Department of Health "Storage Tank Facility Owner/Operator Guidance Document". Regulatory limits are based on RCRA "Maximum Concentration of Contaminants for the Toxicity Characteristic".
[f] The sample was not analyzed for that parameter.
[g] This limit is based on regulations specified in the Colorado Department of Health "Storage Tank Facility Owner/Operator Guidance Document".
[h] No individual regulatory level for this parameter, it is combined as Total BTEX.

*Value outside of QA limits

mg/kg = milligram per kilogram
µg/kg = microgram per kilogram
mg/ℓ = milligram per liter
µg/ℓ = microgram per liter

Table 5 3
Sediment Sampling Results for Rulison Drilling Effluent Pond
(Page 1 of 2)

Compound	Regulatory Limit	Analytical Results for Sampling Conducted in April 1995																	
		Sediment Samples																	
		SD-14	Q[a]	SD-15	Q	SD-16	Q	SD-17	Q	SD-18	Q	SD-19[b]	Q	ST-01	Q	WFR-04[c]	Q	WFR-04[c]	Q
Total Metals (mg/kg)																Dissolved Metals µg/ℓ		Total Metals µg/ℓ	
Aluminum		14,200		11,400		2,540		14,900		3,250		40		Na[g]		47.5	B	58.5	B
Antimony		0.38	B[a]	0.32	U[a]	0.32	U	0.32	U	0.32	U	0.32	U	NA		9.4	B	1.6	U
Arsenic	100[d]	9.9		7.6		0.41	B	12.1		0.52	B	0.72	B	NA		1.1	U	1.1	U
Barium	2,000[d]	219		161		128		195		136		113		NA		3.4	B	3.8	B
Beryllium		0.86	B	0.72	B	0.32	B	0.93	B	0.48	B	0.31	B	NA		0.90	U	0.90	U
Cadmium	20[d]	0.52	U	0.67	B	0.52	U	0.52	U	0.52	U	0.52	U	NA		2.6	U	2.6	U
Calcium		4,150		4,940		5,780		13,600		8,390		5,710		NA		2050	B	1790	B
Chromium	100[d]	30.6		22.4		298		26.8		34.5		307		NA		5.0	U	5.0	U
Cobalt		8.3	B	6.7	B	1.4	B	8.0	B	2.4	B	1.3	B	NA		4.4	U	4.4	U
Copper		20.4		17.2		10.8		29.6		11.2		8.5		NA		7.3	U	9.8	U
Iron		20,000		16,000		4,240		18,900		3,240		3,350		NA		65.3	B	38.9	U
Lead	100[d]	13.6		11.9		9.2		12.4		8.5		8.7		NA		1.1	U	1.1	U
Magnesium		5,880		4,890		1,780		7,220		2,550		1,580		NA		359	B	367	B
Manganese		416		430		94.5		374		128		79.9		NA		5.3	B	2.1	U
Mercury	4[d]	0.05	U	0.05	U	0.05	U	0.05	U	0.05	U	0.05	U	NA		0.10	U	0.10	U
Nickel		27.6		17.5		4.8	B	26.2		8.4		4.6	B	NA		7.9	U	15.4	U
Potassium		1,390		1,740		1,030		1,980		960	B	662	B	NA		347	U	1180	U
Selenium	20[d]	0.22	U	0.22	U	0.22	U	0.22	U	0.22	U	0.22	U	NA		1.1	U	1.1	U
Silver	100[d]	1.1	U	1.1	U	1.1	U	1.1	U	1.1	U	1.1	U	NA		5.5	U	5.5	U
Sodium		275	B	426	B	1,120		576	B	1,110		690	B	NA		728	B	397	B
Thallium		0.37	B	0.22	U	0.22	U	0.22	U	0.22	U	0.22	U	NA		1.1	U	1.1	U
Vanadium		35.7		26.4		5.0	B	32.5		7.9	B	5.3	B	NA		15.5	U	15.5	U
Zinc		50.2		51.2		22.7		57.4		26.2		18.1		NA		85.6		13.8	B

Refer to footnotes at end of table.

Table 5-3
Sediment Sampling Results for Rulison Drilling Effluent Pond
(Page 2 of 2)

Compound	Regulatory Limit	Analytical Results for Sampling Conducted in April 1995																	
		Sediment Samples																	
		SD-14	Q[a]	SD-15	Q	SD-16	Q	SD-17	Q	SD-18	Q	SD-19[b]	Q	ST-01	Q	WFR-04[c]	Q	WFR-04[c]	Q
TCLP Metals mg/ℓ																			
Arsenic	5.0[e]	0.035	U	0.035	U	0.035	U	0.035	U	0.035	U	0.035	U	NA		NA		NA	
Barium	100[e]	0.76		0.49		0.92		0.44		0.88		1.1		NA		NA		NA	
Cadmium	1.0[e]	0.0024	U	0.0024	U	0.0024	U	0.0024	U	0.0024	U	0.0024	U	NA		NA		NA	
Chromium	5.0[e]	0.0080	B	0.0047	U	0.23		0.0047	U	0.026		0.17		NA		NA		NA	
Lead	5.0[e]	0.028	U	0.028	U	0.029	B	0.028	U	0.028	U	0.028	U	NA		NA		NA	
Mercury	0.2[e]	0.00019	B	0.00010	U	0.00010	U	0.00010	U	0.00010	U	0.00020		NA		NA		NA	
Selenium	1.0[e]	0.038	U	0.038	U	0.38	U	0.038	U	0.038	U	0.038	U	NA		NA		NA	
Silver	5.0[e]	0.0047	B	0.0041	U	0.0041	U	0.0041	U	0.0041	U	0.0041	U	NA		NA		NA	
TPH mg/kg																			
Diesel	[f]	NA		NA		NA		NA		NA		NA		25	U	NA		NA	
Waste Oil	[f]	NA		NA		NA		NA		NA		NA		25	U	NA		NA	
BTEX mg/kg																μg/ℓ			
Benzene	[h]	NA		NA		NA		NA		NA		NA		2.0	U	2.0	U	2.0	U
Toluene	[h]	NA		NA		NA		NA		NA		NA		2.0	U	2.0	U	2.0	U
Ethylbenzene	[h]	NA		NA		NA		NA		NA		NA		2.0	U	2.0	U	2.0	U
Xylene	[h]	NA		NA		NA		NA		NA		NA		4.0	U	4.0	U	4.0	U

[a] Q = Laboratory assigned data qualifier: U = Compound was analyzed for but not detected; B = In organics, the analyte was found in the blank. In inorganics, the result is above the Instrument Detection Limit but below the Contract Required Detection Limit.
[b] Duplicate sample of SD-18
[c] Quality Assurance Sample
[d] No regulations for these soil parameters are specified in the Colorado Department of Health "Storage Tank Facility Owner/Operator Guidance Document". Regulatory limits are based on 20X the RCRA "Maximum Concentration of Contaminants for the Toxicity Characteristic".
[e] No regulations for these soil parameters are specified in the Colorado Department of Health "Storage Tank Facility Owner/Operator Guidance Document". Regulatory limits are based on RCRA "Maximum Concentration of Contaminants for the Toxicity Characteristic".
[f] Regulatory limits for these parameters specified in the Colorado Department of Health "Storage Tank Facility Owner/Operator Guidance Document". Regulatory limits are based on RCRA "Maximum Concentration of Contaminants for the Toxicity Characteristic".
[g] The sample was not analyzed for that parameter.
[h] No individual regulatory level for this parameter, it is combined as Total BTEX.

mg/kg = milligram per kilogram
μg/kg = microgram per kilogram
mg/ℓ = milligram per liter
μg/ℓ = microgram per liter

Table 5-4
Soil Sampling Results for Rulison Drilling Effluent Pond
(Page 1 of 3)

Compound	Regulatory Limit	Soil Samples Collected Sept. and Oct. 1994						Soil Samples Collected in April 1995						Field Rinsate	
		SS-01	Q[a]	SS-02	Q	SS-03	Q	SS-04	Q	SS-05[b]	Q	SS-06	Q	WFR-03[c]	Q
Total Metals (mg/kg)														µg/ℓ	
Aluminum		7,300		6,320		13,000		5,940		3,710		11,000		26.2	B
Antimony		0.36	U[a]	0.36	U	0.45	B[a]	0.32	U	0.32	U	0.32	U	1.8	U
Arsenic	100[d]	5.4		2.4		15		2.6		2	B	6.2		1	U
Barium	2,000[d]	**2,530**		**6,040**		206		**6,870**		**5,000**		895		2.1	B
Beryllium		0.65	B	0.71	B	0.82	B	0.63	B	0.55	B	0.71	B	0.2	U
Cadmium	20[d]	0.74	U	0.74	U	0.86	B	0.71	B	0.52	U	0.52	U	3.7	U
Calcium		4,950		10,600		6,270		6,950		7,130		4,640		115	B
Chromium	100[d]	**467**		**857**		25.5		**779**		**750**		**112**		3.4	U
Cobalt		6.5	B	9.3	B	9.2	B	10.2		7.4	B	7.4	B	3.2	U
Copper		19.7		26.1		18.5		23.4		21.5		15.2		9.3	B
Iron		12,300		12,500		20,200		11,500		9,250		16,200		110	
Lead	100[d]	47.6		84		18.6		86.3		77.8		18.3		1	U
Magnesium		3,230		3,550		6,540		2,920		2,220		4,040		113	B
Manganese		294		279		445		286		218		272		1.8	B
Mercury	4[d]	0.05	U	0.06	B	0.05	B	0.05	U	0.05	U	0.05	U	0.16	B
Nickel		14.4		11.3		19.4		11.5		9.1		16.1		15.5	U
Potassium		1,560		2,400		1,560		1,730		1,400		2,260		1,940	U
Selenium	20[d]	0.26	U	0.26	U	0.26	U	0.22	U	0.22	U	0.22	U	1.3	U
Silver	100[d]	0.74	U	0.74	U	0.74	U	1.1	U	1.1	U	1.1	U	3.7	U
Sodium		2,020		1,080		774	B	208	B	279	B	109	B	250	B
Thallium		0.3	U	0.3	U	0.33	B	0.22	U	0.22	U	0.22	U	1.5	U
Vanadium		14.2		9.4	B	36.3		10		5.9	B	22.3		5.6	B
Zinc		135		245		54.1		243		221		67.6		19.7	B

Refer to footnotes at end of table.

Table 5-4
Soil Sampling Results for Rulison Drilling Effluent Pond
(Page 2 of 3)

Compound	Regulatory Limit	Soil Samples Collected Sept. and Oct. 1994						Soil Samples Collected in April 1995						Field Rinsate	
		SS-01	Q[a]	SS-02	Q	SS-03	Q	SS-04	Q	SS-05[b]	Q	SS-06	Q	WFR-03[c]	Q
TCLP Metals (mg/ℓ)															
Arsenic	5[e]	NA[f]		NA		NA		0.035	U	0.035	U	0.035	U	NA	
Barium	10[e]	NA		NA		NA		1		0.87		0.62		NA	
Cadmium	1[e]	NA		NA		NA		0.0024	U	0.0024	U	0.002	U	NA	
Chromium	5[e]	NA		0.22		NA		0.05		0.12		0.005	B	NA	
Lead	5[e]	NA		0.042	U	NA		0.028	U	0.039	B	0.028	U	NA	
Selenium	1[e]	NA		NA		NA		0.038	U	0.038	U	0.038	U	NA	
Silver	5[e]	NA		NA		NA		0.0041	U	0.0041	U	0.004	U	NA	
Mercury	0.20[e]	NA		NA		NA		0.00010	U	0.00014	B	0.0001	U	NA	
TPH (mg/kg)														ug/ℓ	
Nonspecific	250[g]	NA		NA		NA		NA		NA		NA		NA	
Gas		16		75	•	0.83	•	NA		NA		NA		100	U
Diesel		12,000		73,000		25	U	NA		NA		NA		500	U
Waste Oil		250	U	2,500	U	54		NA		NA		NA		500	U

Refer to footnotes at end of table.

Table 5-4
Soil Sampling Results for Rulison Drilling Effluent Pond
(Page 3 of 3)

Compound	Regulatory Limit	Soil Samples Collected Sept. and Oct. 1994						Soil Samples Collected in April 1995						Field Rinsate	
		SS-01	Q[a]	SS-02	Q	SS-03	Q	SS-04	Q	SS-05[b]	Q	SS-06	Q	WFR-03[c]	Q
BTEX (μg/kg)														ug/ℓ	
Benzene	h	4.9		38		2	U	NA		NA		NA		NA	
Toluene	h	17		570		2	U	NA		NA		NA		NA	
Ethylbenzene	h	120		570		2	U	NA		NA		NA		NA	
Xylene	h	500		2,800		2	U	NA		NA		NA		NA	
Total BTEX	50,000	641.9		3,978		2	U	NA		NA		NA		NA	

[a] Q = Laboratory assigned data qualifier: U = Compound was analyzed but not detected; B = In organics, the analyte was found in the blank. In inorganics, the result is above the Instrument Detection Limit but below the Contract Required Detection Limit.
[b] Duplicate of Sample SS-04
[c] Field Rinsate taken during October 1994 sampling event
[d] No regulations for these soil parameters are specified in the Colorado Department of Health "Storage Tank Facility Owner/Operator Guidance Document." Regulatory limits are based on 20X RCRA Maximum Concentration of Contaminants for the Toxicity Characteristic."
[e] No regulations for these soil parameters are specified in the Colorado Department of Health "Storage Tank Facility Owner/Operator Guidance Document." Regulatory limits are based on RCRA Maximum Concentration of Contaminants for the Toxicity Characteristic."
[f] The sample was not analyzed for that parameter.
[g] This limit is based on regulations specified in the Colorado Department of Health "Storage Tank Facility Owner/Operator Guidance Document."
[h] No individual regulatory level for this parameter, it is combined as Total BTEX.

* Value outside of QA limits

mg/kg = milligram per kilogram
mg/ℓ = milligram per liter
μg/kg = microgram per kilogram
μg/ℓ = microgram per liter

Table 5-5
Surface Water Sampling Results for Rulison Drilling Effluent Pond
(Page 1 of 2)

Compound	Regulatory Limit[a]	Sampling Conducted in Sept. and Oct. 1994								Sampling Conducted in April 1995								Sampling Conducted in April 1995								October 1994		April 1995	
		Pond Surface Water Samples								Stream		Spring		Stream		Spring		Pond Surface Water Samples								Rinsate Samples			
		SW-01	Q[b]	SW-02	Q	SW-03	Q	SW-04	Q	SWST-01	Q	SWS-01	Q	SWST-01	Q	SWS-01	Q	SWP-01	Q	SWP-02	Q	SWP-03[e]	Q	SWP-04	Q	WFR-01	Q	WF-04	Q
Total Metals (µg/l)										Total Metals (µg/l)				Dissolved Metals (µg/l)															
Aluminum		52.4	B[b]	135	B	43.2	B	77.2	B	228		32.5	U	34.1	B	55.6	B	NA[c]		NA		NA		NA		27.5	B	NA	
Antimony		1.8	U[b]	1.8	U	1.8	U	1.8	U	1.6	U	3.3	B	3.2	B	6	B	NA		NA		NA		NA		1.8	U	NA	
Arsenic	50	7.4	B	7.5	B	7	B	7.4	B	1.3	B	1.1	U	1.1	U	1.1	U	NA		NA		NA		NA		1	U	NA	
Barium	1,000	51.2	B	52.8	B	49.8	B	51.9	B	52.8	B	46.4	B	47.9	B	45.8	B	NA		NA		NA		NA		1.8	B	NA	
Beryllium		0.86	B	0.29	B	0.29	B	0.29	B	1.7	B	0.9	U	0.9	U	1.4	B	NA		NA		NA		NA		0.21	B	NA	
Cadmium	10	3.7	U	3.7	U	3.7	U	3.7	U	2.6	B	2.6	U	2.6	U	2.6	U	NA		NA		NA		NA		3.7	U	NA	
Calcium		23,400		23,800		22,900		23,500		43,300		83,100		42,600		82,000		NA		NA		NA		NA		90.4	B	NA	
Chromium	50	3.4	U	3.4	U	3.4	U	3.9	B	5	U	5	U	5	U	5	U	NA		NA		NA		NA		3.4	U	NA	
Cobalt		3.2	U	3.2	U	3.2	U	3.2	U	4.4	U	4.4	U	4.4	U	4.4	U	NA		NA		NA		NA		3.2	U	NA	
Copper		10.7	B	12.5	B	8.5	B	16.9	B	10.6	B	9.8	U	7.3	U	9.6	B	NA		NA		NA		NA		9.3	B	NA	
Iron		62.2	B	201		61	B	177		239		38.9	U	38.9	U	38.9	U	NA		NA		NA		NA		46	B	NA	
Lead	50	1	U	1	U	1	U	1	U	1.1	U	1.1	U	1.1	U	1.1	U	NA		NA		NA		NA		1	U	NA	
Magnesium		28,900		29,200		28,700		29,400		15,300		58,100		15,200		56,400		NA		NA		NA		NA		135	B	NA	
Manganese		6	B	18.9		5.5	B	8.3	B	12.9	B	2.1	U	2.7	B	2.1	U	NA		NA		NA		NA		2.9	B	NA	
Mercury	2	0.14	B	0.16	B	0.16	B	0.16	B	0.1	U	0.1	U	0.1	U	0.1	U	NA		NA		NA		NA		0.16	B	NA	
Nickel		15.5	U	15.5	U	15.5	U	15.5	U	15.4	U	15.4	U	7.9	U	7.9	U	NA		NA		NA		NA		15.5	U	NA	
Potassium		2,030	B	1,940	U	1,940	U	1,940	U	1,860	B	1,180	U	1,890	B	1,240	B	NA		NA		NA		NA		1,940	U	NA	
Selenium	10	6.5	U	6.5	U	6.5	U	6.5	U	1.5	B	2.1	B	1.1	U	2.7	B	NA		NA		NA		NA		1.3	U	NA	
Silver	50	3.7	U	3.7	U	3.7	U	3.7	U	5.5	U	5.5	U	5.5	U	5.5	U	NA		NA		NA		NA		3.7	U	NA	
Sodium		51,400		52,500		51,700		52,300		18,900		52,800		19,900		50,600		NA		NA		NA		NA		183	B	NA	
Thallium		1.5	U	1.5	U	1.5	U	1.5	U	1.1	U	1.1	U	1.1	U	1.1	U	NA		NA		NA		NA		1.5	U	NA	
Vanadium		11	B	11	B	9.7	B	10.5	B	15.5	U	15.5	U	15.5	U	15.5	U	NA		NA		NA		NA		4	B	NA	
Zinc		9.4	B	14.2	B	11.1	B	11.3	B	13.2	B	10.3	B	10.2	B	9.6	B	NA		NA		NA		NA		7.3	B	NA	
TPH[d] (mg/l)																													
Nonspecific		0.48	U	0.51	U	0.48	U	0.56	U	NA		NA		NA		NA		NA		NA		NA		NA		0.49	U	NA	
Gas		NA		NA		NA		NA		NA		NA		NA		NA		NA		NA		NA		NA		NA		NA	
Diesel		NA		NA		NA		NA		NA		NA		NA		NA		NA		NA		NA		NA		NA		NA	
Waste Oil		NA		NA		NA		NA		NA		NA		NA		NA		NA		NA		NA		NA		NA		NA	

Refer to footnotes at end of table.

Table 5-5
Surface Water Sampling Results for Rulison Drilling Effluent Pond
(Page 2 of 2)

Compound	Regulatory Limit[a]	Sampling Conducted in Sept. and Oct. 1994								Sampling Conducted in April 1995								Sampling Conducted in April 1995								October 1994		April 1995	
		Pond Surface Water Samples								Stream		Spring		Stream		Spring		Pond Surface Water Samples								Rinsate Samples			
		SW-01	Q[b]	SW-02	Q	SW-03	Q	SW-04	Q	SWST-01	Q	SWS-01	Q	SWST-01	Q	SWS-01	Q	SWP-01	Q	SWP-02	Q	SWP-03*	Q	SWP-04	Q	WFR-01	Q	WF-04	Q
Tritium (pCi/l)														**	**	**	**												
Tritium		-2		40		78		70		NA		NA		**	**	**	**	NA		NA		NA		NA		-2		NA	
2 Sigma Error (+/-)		103		105		106		106		NA		NA		**	**	**	**	NA		NA		NA		NA		103		NA	
MDA		178		178		178		178		NA		NA		**	**	**	**	NA		NA		NA		NA		178		NA	
BTEX[e] (ug/kg)																													
Benzene	5[f]	NA		NA		NA		NA		2	U	2	U	**	**	**	**	2	U	2	U	2	U	2	U	NA		2	U
Toluene	1,000	NA		NA		NA		NA		2	U	2	U	**	**	**	**	2	U	2	U	2	U	2	U	NA		2	U
Ethylbenzene	680	NA		NA		NA		NA		2	U	2	U	**	**	**	**	2	U	2	U	2	U	2	U	NA		2	U
Xylene	10,000	NA		NA		NA		NA		4	U	4	U	**	**	**	**	4	U	4	U	4	U	4	U	NA		4	U

[a]No regulations for metal concentrations, Total Petroleum Hydrocarbon and Radionuclides are specified in the Colorado Department of Health "Storage Tank Facility Owner/Operator Guidance Document." Colorado water clean-up standards are site specific and based on "The Basic Standards for Groundwater (5CCR1002-8)." The Safe Drinking Water Standards have been provided for comparison purposes only.
[b]Q = Laboratory assigned data qualifier: U = Compound was analyzed for but not detected; B = In inorganics, the result is above the Instrument Detection Limit but below the Contract Required Detection Limit.
[c]The sample was not analyzed for that parameter.
[d]Total Petroleum Hydrocarbons
[e]Benzene, Toluene, Ethylbenzene, Xylene
[f]This limit is based on regulation specified in the Colorado Department of Health "Storage Tank Facility Owner/Operator Guidance Documents."

*Duplicate of sample SWP-02
**These samples were not analyzed for these parameters.

mg/kg = milligram per kilogram
μg/kg = microgram per kilogram
μg/ℓ = microgram per liter

Table 5-6
Fish Sampling Results for Rulison Drilling Effluent Pond

Compound	Sampling Conducted in September and October 1994							
	F-01	Q[a]	F-02	Q	F-03	Q	WFB-01[b]	Q
Metals (mg/kg)							mg/ℓ	
Aluminum	6.40	B[a]	5	U[a]	5	U	59.9	B
Antimony	0.36	U	0.36	U	0.36	U	1.8	U
Arsenic	0.26	U	0.26	U	0.43	B	1.0	U
Barium	0.31	B	0.21	B	0.16	B	1.4	B
Beryllium	0.04	B	0.04	U	0.04	U	0.29	B
Cadmium	0.74	U	0.74	U	0.74	U	3.7	U
Calcium	369	B	426	B	453	B	69.2	B
Chromium	0.68	U	0.68	U	0.68	U	3.4	U
Cobalt	0.64	U	0.64	U	0.64	U	3.2	U
Copper	0.77	B	0.5	U	0.5	U	8.4	B
Iron	12.6	B	11.1	B	5.7	B	71.0	B
Lead	0.2	U	2	U	2	U	1.0	U
Magnesium	259	B	233	B	284	B	155	B
Manganese	0.14	U	0.14	B	0.14	U	1.7	B
Mercury	0.13		0.[illegible]5	B	0.08	B	0.15	B
Nickel	3.1	U	3.1	U	3.1	U	15.5	U
Potassium	4,870		3,880		4,490		1,940	U
Selenium	0.26	U	0.26	U	0.26	U	1.3	U
Silver	0.74	U	0.74	U	0.74	U	3.7	U
Sodium	609	B	693	B	780	B	168	B
Sodium	0.3	U	0.3	U	0.3	U	1.5	U
Vanadium	0.62	U	0.62	U	0.62	U	4.9	B
Zinc	6.2		6.3		9.1		13.5	B
TPH (mg/kg)[c]								
Nonspecific	13.7		31.5		17.3		0.49	U
Gas	NA[d]		NA		NA		NA	
Diesel	NA		NA		NA		NA	
Waste Oil	NA		NA		NA		NA	

[a]Q = Laboratory assigned data qualifier: U = Compound was analyzed for but not detected; B = In inorganics, the result is above the Instrument Detection Limit but below the Contract Required Detection Limit.
[b]Field blank sample
[c]Total Petroleum Hydrocarbons
[d]The sample was not analyzed for that parameter.

mg/kg = milligram per kilogram
mg/ℓ = milligram per liter

Sample population size has been determined through the use of power curves, presented in the 1989 EPA guidance document "Methods for Evaluating the Attainment of Cleanup Standards, Volume 1: Soils and Solid Media" (EPA, 1989a) for testing the mean in determining the attainment of clean-up standards. The sample population size was determined for TPH, as this COC has the greatest likelihood of requiring additional samples to demonstrate clean-up attainment. For the purposes of this test, the following assumptions were made:

$$\alpha = 10\%$$
$$\beta = 20\%$$
$$\mu_1 = 0.46 \cdot Cs$$
$$Cs = 250 \text{ milligrams per kilogram TPH}$$
$$cv = 1.29$$

where

α = false positive rate (i.e., site is thought to be clean but is not)

β = false negative rate (i.e., site is thought to be contaminated but is not)

μ_1 = mean analytical result desired is determined from the associated power curve corresponding to the given α and β

Cs = clean-up standard

cv = coefficient of variation for the data as the ratio of the standard deviation to the mean.

Using the power curve "C" for $\alpha = 10\%$, and estimating the cv for the currently existing data at a desired mean of 46 percent of the Colorado clean-up standard for TPH, the estimated number (n_d) of samples to be collected to determine if clean-up standards have been met is 35 TPH samples.

Using the number of samples obtained above, the final sample size is determined by the following:

$$n_f = n_d / (1 - R)$$

where

n_d = estimated sample size (35)

R = rate that missing or unusable data will occur (1% unusable)

n_f = final sample size

Thus, the total number of TPH samples to collect is estimated at 36.

The sample grid is calculated from the following equation:

$$L = (A / n_f)^{1/2}$$

where

L	=	distance between sample points (in m)
A	=	total area (in m^2) to be sampled (2,024 m^2)
n_f	=	sample size (36)

Thus, the distance between sampling points on the sampling grid is estimated to be 7.5 m. Based on a total of 36 TPH samples, 36 BTEX samples and 8 total RCRA metals samples will be collected. Total metals sample locations will be chosen randomly from the nodes on the grid, using a random numbers table. An additional verification soil sample will be collected from the settling area located on the west side of the drilling effluent pond. A diagram showing the proposed layout for verification soil samples is included as Figure 5-1.

5.6 Sampling and Sample Handling Procedures

Prior to beginning sampling activities, all required permits and/or written approvals will be obtained and all required materials and equipment will be staged at the site. The Environmental Services Support Contractor shall verbally notify the persons granting authorization to sample the site prior to sampling activities.

An SSHASP shall also be prepared and approved prior to initiating sampling activities. All sampling activities will be conducted in accordance with an approved SSHASP. Any basic protective clothing or equipment required for sampling will be specified in the SSHASP. Procedures and requirements for sample collection, preservation, handling, and analysis are detailed below.

5.7 Sample Collection

Sampling will be conducted in accordance with approved contractor procedures. A sampling grid will be constructed at the sampling sites based on the calculations presented in Section 5.1.4 of this VSAP. One TPH and BTEX environmental sample will be collected at each of the sampling nodes on the grid. Eight total RCRA metals samples will be collected from randomly chosen nodes on the sampling grid. Additionally, one MS/MSD sample and one duplicate sample will be collected from randomly chosen locations and intervals along the grid, at rates

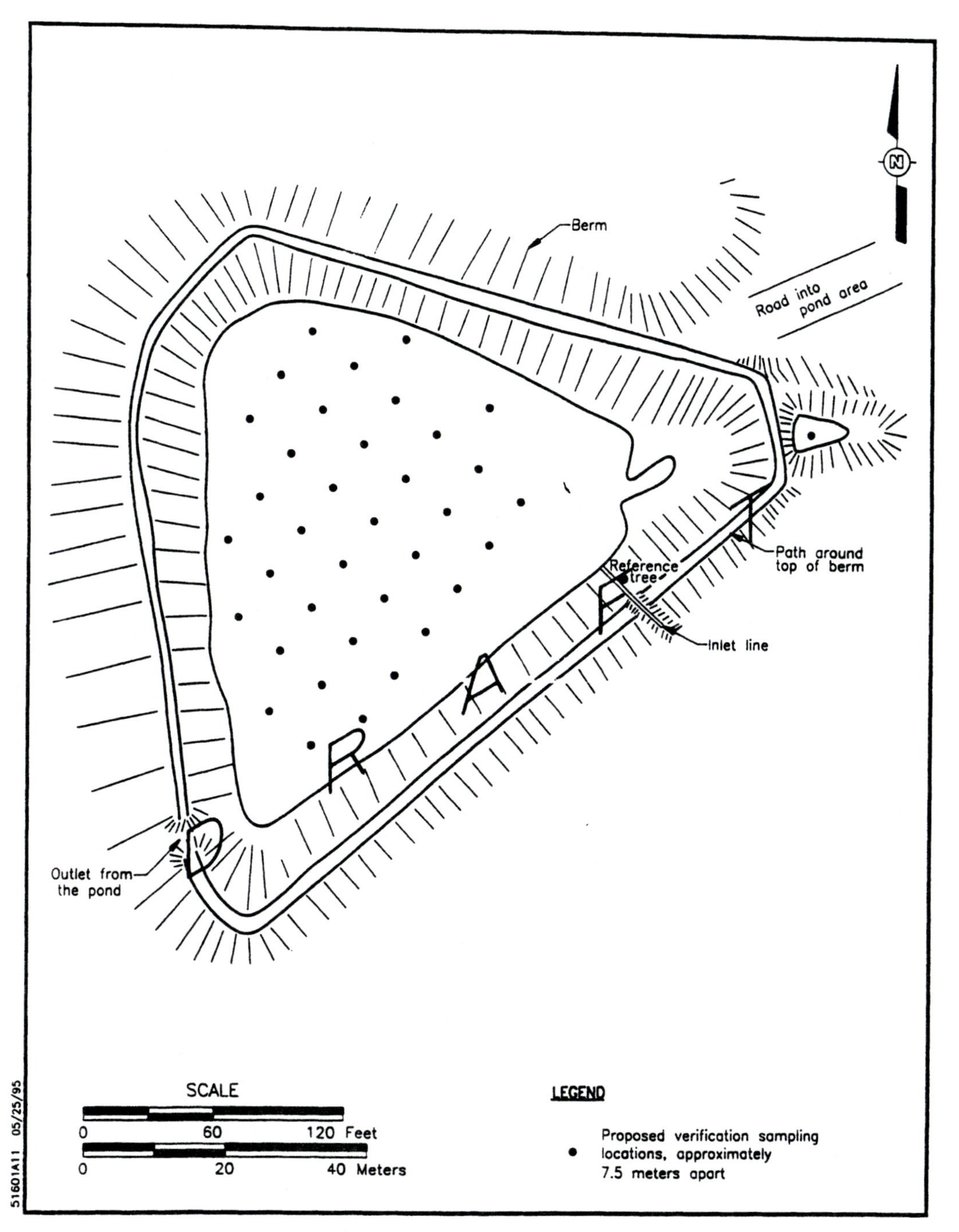

Figure 5-1
Proposed Verification Soil Sampling Locations, Rulison Drilling Effluent Pond

indicated in Chapter 4.0 of this VSAP. The MS/MSD and duplicate samples will not be collected from the same location/interval as one another. The QC samples will be split (collected from) the same volume of sediment from which the associated environmental sample was taken. A detailed geologic description will be made of every environmental sample collected.

To achieve unbiased sample splits, each sample volume will be homogenized in a decontaminated stainless-steel mixing bowl, using a stainless-steel trowel or spatula, prior to splitting the sample fractions. Samples for BTEX will not be homogenized. Equipment rinsate samples will be collected at rates specified in Chapter 4.0 of this VSAP, and source water blank samples will be collected during the execution of this plan. Liquid samples (equipment rinsate and deionized water field blanks) will consist of one 1-liter polyethylene-bottle for metals analysis, and three 40-mℓ glass VOA-vials for BTEX analysis. One 500-mℓ TPH and one 250-mℓ amber glass shall be collected from each soil sample location. One 500-mℓ amber glass for metals analysis will be collected from the 8 randomly-chosen nodes on the grid.

Soil samples shall be collected following the procedures listed below:

- Construct the sampling grid according to the calculations presented above. If no sampling grid is possible because of sediment clean-up operations, each sampling location will be identified and located at the approximate distance form each neighbor sampling location as the cleanup progresses.

- Don clean, dry, surgical gloves prior to beginning sampling activities. Cotton gloves may be worn underneath the surgical gloves. Surgical gloves will be replaced after each discreet sample is collected.

- After sampling activities at a given sampling location are completed, documented, and verified, fill in the hole with excess excavated sediment (if necessary). Prior to sampling the next location, decontaminate the sampling tools according to approved contractor procedures.

- Perform a final decontamination of the sampling and mixing equipment using the same decontamination procedure (Section 5.4). Collect a single equipment rinsate sample by pouring deionized water over the sampling device and collecting the sample directly from the stream of water coming off the device. Also, collect an aliquot of each source of deionized water used for the equipment rinsate sample and the decontamination activities. Containerize the decontamination water for later disposal.

- Prior to leaving the site, survey the area to be sure that it is left in its original condition and that no materials or wastes are left behind.

5.8 Sample Preservation

All samples will be preserved by cooling to approximately 4 degrees Celsius (°C). Samples will be placed in a shipping cooler with frozen cooling gel packs and/or ice at the site. The target temperature of $4°C \pm 2°C$ will be maintained until the laboratory receives the shipment. The temperature of the water shall be assessed upon arrival to the laboratory. Equipment rinsate and deionized-water blank for metals analysis also include a chemical preservative as specified in the Rulison Site QAPP (DOE, 1995b).

5.9 Sample Handling and Analysis

The Rulison Site QAPP (DOE, 1995b) lists analyses, sample containers, preservatives, holding times, quantitation limits, and analytical and laboratory methods expected for this task. Analyses of all metals and organic compounds will be consistent with EPA SW-846 methodologies to provide consistent and reliable data.

All holding times, quantitation limits, critical levels, and decision amounts will be met as outlined in the Rulison Site QAPP (DOE, 1995b). Failure to meet the recommended holding times, quantitation limits, critical levels, and decision amounts may result in qualified or unacceptable data.

5.10 Decontamination Procedures

The analytical laboratory selected will supply sample containers to the field sampling personnel in a precleaned condition. Sample jar cleaning will follow established EPA guidance. In the event that the analytical laboratory cannot supply the required sample containers, containers will be purchased by the ESSC from a qualified supplier who will perform and document the sample container precleaning. Certificates of cleanliness shall be supplied for all precleaned sample containers.

Decontamination of field sampling equipment is required for all sampling tasks. A thorough decontamination of the sampling and sample mixing equipment will be performed between sampling locations at a given site using a clean scrub brush, a laboratory-grade soap (such as Alconox®) wash, isopropanol or nitric acid rinse, and a deionized water rinse. All decontamination activities performed during soil or water sampling shall be performed in accordance with approved contractor procedures.

5.11 Waste and Contaminated Materials Disposal

No investigation-derived waste will be left on site as a result of the sampling activities. Excess soil from the sampling process will be returned to its original location. Equipment rinsate and decontamination solutions will be stored in appropriate waste containers and disposed in an appropriate fashion consistent with waste disposal guidelines.

6.0 Sample Documentation and Custody

Samples will be properly handled in accordance with approved contractor procedures to maintain sample integrity from collection through analysis. The following sections identify actual sample management and custody requirements for the study. Any significant change or nonconformance in technical procedure shall warrant official documentation.

6.1 Field Operations

To document the integrity of samples from the time of collection through data reporting, sample collection and custody records shall be maintained. Standardized forms (including Sample Collection Logs and the AR/COC) will be used to document sample collection and sample custody during field investigations. These forms will be completed using approved contractor procedures. The sampling grid locations will be clearly documented using detailed narrative and field sketches drawn on attachments to the Sample Collection Log. A detailed geologic description of the soil shall be recorded on a Field Activity Daily Log at each sampling location.

All documentation must be legible, identifiable, and recorded in permanent black ink. Field personnel will complete field documentation at the job site during or immediately after sample collection. Errors on forms are corrected by drawing a single line through the error, so that the stricken text remains legible. The correct information will be entered along with the date and initials of the person recording the information. All entries on the forms will be completed. In the event that the entry is not applicable, it will be noted by marking "NA" or by lining out the entry. Each component of the sample control and documentation process is briefly described below.

All sample documentation, including sample labels, sample collection logs, and analysis requests and chain-of-custody records, will be collected in accordance with approved contractor procedures.

6.2 Sample Identification Numbers

Each sample will be uniquely identified with an identification number issued by the ESSC Project Manager. Sample codes for grid locations or water sample collection locations may be recorded on the Sample Collection Log and AR/COC, if necessary.

6.3 Laboratory Operations

Laboratory sample custody, sample analysis, data management, reporting, and sample disposal will be performed in accordance with established laboratory procedures approved by the ESSC Project Manager.

7.0 Analytical Procedures

Analytical procedures will follow established laboratory procedures based on the referenced EPA methods. Analyses, sample containers, preservatives, holding times, quantitation limits, and analytical and laboratory methods to be used in this task are cited in the Rulison Site QAPP (DOE, 1995b).

Instrument calibration, calibration source traceability, analytical QC, and QC acceptance criteria will be in accordance with the contractor laboratory's quality assurance plan, approved by the ESSC Project Manager, and in the contract Statement of Work between the DOE and the laboratory.

8.0 Data Reduction, Validation and Reporting

The assigned contractor analytical laboratory will perform initial data reduction and validation. Data reported by the laboratory will meet method and laboratory QC requirements. The laboratory will analyze duplicate laboratory control samples for indicators of bias and precision and will report results as percent recovery and relative percent difference of the samples and the duplicate samples. The analytical report will include the QC acceptance criteria for bias and precision (see Section 4.2 of this VSAP). The laboratory will provide a summary data report and will archive all raw data, bench sheets, and other relevant information in a retrievable manner until requested by the DOE.

8.1 Measurement Data and Sample Collection Documentation Review

The ESSC will verify sampling and analytical data generated under this VSAP for analytical contract compliance, including review of analyte quantitation (reporting) limits and QC indicators. The ESSC Project Manager will provide documentation of the verification process with transmittal of the verified data package and the final report. Data will be verified as set forth in approved contractor procedures.

8.2 Data Assessment

Following receipt of validated and verified analytical and sampling data, the ESSC Project Manager will assess the analytical results for COC detection criteria.

8.3 Data Reporting

The contractor laboratory will transmit summary analytical and laboratory QC data to the ESSC. The analytical report will be in electronic and hardcopy formats, generated from a single source. The analytical laboratory will archive all raw data, notes, and bench sheets until those records are requested by the DOE. The ESSC Project Manager will transmit all original field and sample custody documentation, verification and validation documentation, and the analytical report to the DOE/NV Rulison Site Manager as part of the report. The ESSC will prepare a report presenting the data in tabulated form to the DOE/NV Rulison Site Manager.

9.0 Quality Reports to Management

The ESSC report will identify areas of concern encountered during project sampling and analysis efforts, as well as possible resolutions in an effort to improve data quality in future similar projects. Additional quality reports to management will include nonconformance and corrective actions and assessment results, if necessary.

10.0 Nonconformances and Corrective Actions

Nonconformances are items or activities that do not meet the project requirements of approved procedures. Unlike variances, which are preapproved and controlled, they are uncontrolled and unapproved deviations. Nonconformances to the activities specified in this VSAP will be documented and evaluated in accordance with the Rulison Site QAPP.

Whenever possible, corrective actions will be applied to rectify or prevent reoccurrence of nonconformances or other conditions that could adversely affect the quality project data. Corrective actions will be implemented in accordance with approved contractor procedures.

11.0 Assessments

Following approved contractor procedures, the ESSC Project and the ESSC Health and Safety Officer will conduct assessments of the sampling activities. Field assessments are used to determine if field procedures are being conducted in compliance with the applicable VSAP and SSHASP. Items reviewed may include, but are not limited to, sample collection and handling, documentation, sampling technique, equipment calibration, maintenance procedures, and health and safety practices.

The contractor's analytical laboratories participate in system audits as part of the procurement selection process. Additionally, the contractor's laboratories are required to participate in external performance audits or evaluation programs sponsored by the EPA or other state accreditation organizations.

12.0 Records Management

Completed records generated during sampling and analysis will be submitted by the ESSC Project Manager to the DOE/NV Rulison Site Manager for archival at the DOE/NV. The DOE/NV Rulison Site Manager responsible for this task will submit all documentation to the DOE/NV records center upon completion of the project.

The laboratory shall retain and make available for inspection upon request all raw analytical records generated in conjunction with this VSAP. These records shall include instrument tuning and calibration records, batch QC sample data, control charts and calculations, sample tracking and control documentation, raw analytical sample data, and analytical results. These records shall be retained for a duration of time specified in the contract Statement of Work until requested by the DOE.

13.0 References

EPA, See U.S. Environmental Protection Agency.

ERDA, See U.S. Energy Research and Development Administration

U.S. Department of Energy, Nevada Operations Office, 1995a, Draft Rulison Corrective Action Plan, IT Corporation, Las Vegas, NV.

U.S. Department of Energy, Nevada Operations Office, 1995b, Draft Rulison Drilling Effluent Pond Remediation Quality Assurance Project Plan, Rev. 0, IT Corporation, Las Vegas, NV.

U.S. Department of Energy, Nevada Operations Office, 1995c, Draft Rulison Site-Specific Health and Safety Plan, IT Corporation, Las Vegas, NV.

U.S. Department of Energy, Nevada Operations Office, 1988.

U.S. Energy Research and Development Administration, 1977.

U.S. Environmental Protection Agency, 1993, *Guidance for Planning for Data Collection in Support of Environmental Decision Making Using the Data Quality Objectives Process,* EPA QA/G-4 Interim Final, U.S. Environmental Protection Agency, Quality Assurance Management Staff, Washington, DC.

U.S. Environmental Protection Agency, 1992, *Statistical Analysis of Ground-Water Monitoring Data at RCRA Facilities, Addendum to Interim Final Guidance,* EPA/530-SW-89-026, U.S. Environmental Protection Agency, Office of Solid Waste, Waste Management Division, Washington, DC.

U.S. Environmental Protection Agency, 1989a, *Methods for Evaluating the Attainment of Cleanup Standards, Volume 1: Soils and Solid Media,* EPA 230/02-89-042, Washington, DC.

U.S. Environmental Protection Agency, 1989b, *Statistical Analysis of Ground-Water Monitoring Data at RCRA Facilities, Interim Final Guidance,* EPA/530-SW-89-026, Washington, DC.

U.S. Environmental Protection Agency, 1989c, *Report on Minimum Criteria to Assure Data Quality,* EPA/530-SW-90-021, U.S. Environmental Protection Agency, Washington, DC.

U.S. Environmental Protection Agency, 1987a, *Data Quality Objectives for Remedial Response Activities, Development Response,* Washington, DC.

U.S. Environmental Protection Agency, 1987b, *Test Methods for Evaluating Solid Waste, Physical/Chemical Methods,* SW-846, 3rd ed., as amended by Update I, 1991, Washington, DC.

U.S. Environmental Protection Agency, 1985, *Verification of PCB Spill Cleanup by Sampling and Analysis,* EPA-560/5-85-026, Washington, DC.

Glossary

AA Atomic absorption spectrophotometer; used in trace metal analysis.

Absorption Collection into a liquid media.

ACGIH American Conference of Governmental Industrial Hygienists.

Adsorption Physical attachment to a surface.

AE Atomic emission.

Aerodynamic diameter Describes the size of an aerosol based on the way that it behaves in an airstream.

Aerosol Solid or liquid particles suspended in a gas.

AES Atomic emission spectroscopy.

Alpha particle Radioactive particle made up of two neutrons and two protons.

Analyte A chemical or contaminant of concern; the chemical for which the sample is being analyzed to determine its presence.

Anthropogenic Human produced.

AOAC Association of Official Analytical Chemists.

APHA American Public Health Association.

Area sample Air sampling in a fixed location or area.

ASTM American Society for Testing and Materials.

Auto ignition temperature Minimum temperature at which a flammable mixture will ignite without spark or flame.

Beta particle Small, negatively charged radioactive particle. Similar in size to an electron.

Breathing zone A sphere 1 foot in radius from the nose and mouth.

BTEX Benzene, toluene, ethylbenzene, and xylenes, constituents of gasoline; of interest in investigations into leaking underground fuel tanks.

CERCLA Comprehensive Environmental Response, Compensation, and Liability Act.

CFR Code of Federal Regulations.

CGI Combustible gas indicator, a device to measure the concentration of a flammable vapor or gas in air, giving results as a percentage of the LEL of the calibration gas.

Chain of custody A procedure used to document the possession of a sample by a given individual during a given time period.

CI Confidence interval.

CLP Contract Laboratory Program, the federal program for procuring analytical laboratory services to support SuperFund enforcement activities.

COLIWASA Composite liquid waste sampler; a device used to collect from containers and vessels a uniform column of liquid that may be composed of several layers or strata.

Combustible liquid Liquid with a flash point >37.8°C and <93.3°C.

Composite sample A sample composed of several equal amounts of sampled material taken at different times or locations.

Cross-contamination Contamination of a sample due to contact with chemical contamnants from another location, sample, or media.

CVAA Cold vapor atomic absorption.

CWA Clean Water Act.

DL Detection limit; may be an actual analytical method detection limit or a reporting limit.

DNAPL Dense nonaqueous phase liquid.

Dust Solid aerosols formed by mechanical action.

EC Electrical conductivity.

ECD Electron capture detector.

EI Electron impact.

EMMI Environmental Monitoring Methods Index.

EMSL Environmental Measurement and Support Laboratories.

EPA Environmental Protection Agency; the agency that protects the environment.

FAA Flame atomic absorption.

FID Flame ionization detector.

Flammable liquid Liquids with a flash point below 37.8°C.

Flammable range The range between UEL and LEL.

Flash point The minimum temperature at which a liquid gives off sufficient vapor to form an ignitable mixture.

FPD Flame photometric detector.

Fume Solid aerosols resulting from recondensation of vapor.

Gamma rays Powerful electromagnetic energy.

Gas A formless fluid that maintains its characteristics at room temperature.

GC Gas chromatograph; used in the analysis of organics.

GC/ECD Gas chromatography with an electron capture detector.

GC/FPD A gas chromatograph equipped with a flame photometric detector.

GFAA Graphic atomic absorption.

GPC Gel permeation chromatography.

Grab sample One sample taken in one container from one discrete location at one interval in time. See *short-term sample.*

Gravimetric sample A preweighing and post-weighing sample filter used to get the total weight of aerosol collected over the sampling period.

HDPE High-density polyethylene; the grade of plastic required in EPA protocol for trace metals sample containers.

Holding time A time limit specified in analytical laboratory procedure for which the sample may be held waiting for analysis in the laboratory.

HPLC High-performance liquid chromatography.

IACS International Annealed Copper Standard.

IAQ Indoor air quality.

ICP See *ICP–AES.*

ICP–AES Inductively coupled plasma-atomic emission spectrometer.

IDLH Immediately dangerous to life or health.

Industrial hygiene The profession devoted to anticipating, recognizing, evaluating, and controlling hazards arising in or from industry.

Inspirable particulate Aerosols ≤100 μm in aerodynamic diameter.

Integrated sample See *long-term sample.*

Intrinsically safe Instruments engineered to use safely in potentially combustible atmospheres.

IQI Image quality indicator.

IR Infrared.

LEL Lower explosive limit.

LFL Lower flammability limit. See also *LEL.*

LNAPL Light nonaqueous phase liquid.

Long-Term sample A sample collected over a period of time to assess average concentrations.

Matrix The media that the sample is in (i.e., water, soil, sludge).

MCL Maximum contaminant level.

Mist Liquid droplet aerosols.

MSDS Material Safety Data Sheet.

NAPL Nonaqueous phase liquid.

NDT Nondestructive testing.

NIOSH National Institute of Occupational Safety and Health.

NIST National Institute of Standards and Technology, formerly the National Bureau of Standards.

NPD Nitrogen phosphorous detector.

NPDES National Pollutant Discharge System; the EPA permit system established to require permitees to monitor and report results of analysis of discharged waters from in-plant processes.

NRC National Regulatory Commission.

NTIS National Technical Information Service.

NTP Normal temperature (25°Centigrade) and pressure (one atmosphere, 760 mm/Hg).

NTU Nephelometric turbidity units, the measurement of turbidity is based on a comparison of the intensity of light scattered by the sample under defined conditions with the intensity of light scattered by a standard reference suspension.

NWWA National Water Well Association.

OSHA Occupational Safety and Health Administration.

PAH Polynuclear atomic hydrocarbon.

PARCC An acronym for the five descriptors commonly used to assess laboratory quality: precision, accuracy, representativeness, completeness, comparability.

Partitioning The interaction of several phases of a sample with each other by transferring contaminants from one phase to one of the other phases.

PE Polyethylene.

PELs Permissible exposure levels, as defined by OSHA.

PCB See *polychlorinated biphenyl.*

Personal sample Sampling to assess the exposure of an individual.

PM_{10} Particulate matter less than 10 μm in aerodynamic diameter.

Polychlorinated biphenyl A chemical substance formerly used in oils and dielectric fluids as a fire retardant.

P.P. Priority pollutants listed by EPA as pollutants of concern.

PPE Personal protective equipment; also polypropylene plastic.

ppb Parts per billion.

ppm Parts per million.

ppt Parts per trillion.

PVC Polyvinyl chloride.

QA Quality assurance.

QAPjP Quality assurance project plan.

QAPP Quality assurance program plan.

Quality assurance The use of procedures designed to provide verification of the accuracy of final sampling effort results.

QC Quality control.

Quantitative sampling Determines airborne concentration.

RCRA Resource Conservation and Recovery Act.

RELs Recommended exposure levels, as defined by NIOSH.

Respirable particulate Aerosols <4 μm in aerodynamic diameter.

RPD Relative percent difference.

Sampling plan A plan used to document the type and number of samples, as well as the proper sample collection procedures and other planning information necessary for conducting a quality sampling effort.

Short-Term sampling Performed over a short period of time (usually less than five minutes).

Site Safety and Health Plan A plan required for sampling work or other physical work on a contaminated site; contains procedures for employees to follow when entering and working on the site.

Smoke Solid or liquid aerosols resulting from incomplete combustion of carbon materials.

Standard methods Reliable air-sampling methods developed by recognized organizations.

SDWA Safe Drinking Water Act.

TCD Thermal conductivity detector.

TCLP Toxicity characteristic leaching procedure.

TEGD *Technical Enforcement Guidance Document.*

Thoracic particulate Aerosols <10 μm in aerodynamic diameter.

TLVs Threshold limit values, as defined by the American Conference of Governmental Industrial Hygienists.

TPH Total petroleum hydrocarbon.

TTLC Total threshold limit concentration, a term used in the State of California's waste regulations (Title 22) in making decisions regarding hazardous waster disposal.

Turnaround time The time that the laboratory takes to analyze the sample and report back to the customer.

TWA Time-weighted average.

UEL Upper explosive limit.

USGS U.S. Geological Society.

UST Underground storage tank.

Vapor Formless fluid resulting from the evaporation of liquids.

Vapor pressure Pressure exerted by a vapor in equilibrium with its condensed phase.

VOA Volative organic analyte/analysis, a term used to describe the analysis of volatile organic compounds by gas chromatography/mass spectrometry.

VOC Volative organic compound.

Volatile contaminants Contaminants subject to evaporation into the atmosphere.

WCOT Wall-coated open tubular.

Well purging The process of removing accumulated static water from the well so that a fresh quantity of groundwater may flow into the well for sample collection.

Index

ISBN 0-02-389534-9
90000
9 780023 895340